Daniel de Graaf

Adherence to and invasion of epithelial cells by Neisseria gonorrhoeae

Daniel de Graaf

Adherence to and invasion of epithelial cells by Neisseria gonorrhoeae

Südwestdeutscher Verlag für Hochschulschriften

Impressum / Imprint

Bibliografische Information der Deutschen Nationalbibliothek: Die Deutsche Nationalbibliothek verzeichnet diese Publikation in der Deutschen Nationalbibliografie; detaillierte bibliografische Daten sind im Internet über http://dnb.d-nb.de abrufbar.

Alle in diesem Buch genannten Marken und Produktnamen unterliegen warenzeichen-, marken- oder patentrechtlichem Schutz bzw. sind Warenzeichen oder eingetragene Warenzeichen der jeweiligen Inhaber. Die Wiedergabe von Marken, Produktnamen, Gebrauchsnamen, Handelsnamen, Warenbezeichnungen u.s.w. in diesem Werk berechtigt auch ohne besondere Kennzeichnung nicht zu der Annahme, dass solche Namen im Sinne der Warenzeichen- und Markenschutzgesetzgebung als frei zu betrachten wären und daher von jedermann benutzt werden dürften.

Bibliographic information published by the Deutsche Nationalbibliothek: The Deutsche Nationalbibliothek lists this publication in the Deutsche Nationalbibliografie; detailed bibliographic data are available in the Internet at http://dnb.d-nb.de.

Any brand names and product names mentioned in this book are subject to trademark, brand or patent protection and are trademarks or registered trademarks of their respective holders. The use of brand names, product names, common names, trade names, product descriptions etc. even without a particular marking in this works is in no way to be construed to mean that such names may be regarded as unrestricted in respect of trademark and brand protection legislation and could thus be used by anyone.

Coverbild / Cover image: www.ingimage.com

Verlag / Publisher:
Südwestdeutscher Verlag für Hochschulschriften
ist ein Imprint der / is a trademark of
AV Akademikerverlag GmbH & Co. KG
Heinrich-Böcking-Str. 6-8, 66121 Saarbrücken, Deutschland / Germany
Email: info@svh-verlag.de

Herstellung: siehe letzte Seite /
Printed at: see last page
ISBN: 978-3-8381-3630-1

Zugl. / Approved by: Berlin, Freie Universität, Diss., 2009

Copyright © 2013 AV Akademikerverlag GmbH & Co. KG
Alle Rechte vorbehalten. / All rights reserved. Saarbrücken 2013

Dedicated to my family

The greatest obstacle to discovery is not ignorance – it is the illusion of knowledge.

Daniel J. Boorstin

Table of contents

Table of contents		3
1.1	Abstract	7
1.2	Zusammenfassung	8
2	**Introduction**	**10**
2.1	Bacteria – host cell interactions	10
2.1.1	Bacterial adherence to and invasion of target cells	10
2.2	Neisseria spp.	12
2.2.1	Meningococcal pathogenesis	12
2.2.2	Gonococcal pathogenesis and clinical manifestations	13
2.2.3	*N. gonorrhoeae* adhesion to and invasion of epithelial cells	14
2.2.3.1	Type IV pili	14
2.2.3.2	Opa proteins	17
2.2.3.3	Lipooligosaccharides (LOS)	19
2.2.3.4	Porin	20
2.2.3.5	Models for *N. gonorrhoeae* invasion	21
2.3	Caveolae and lipid rafts	23
2.3.1	Biochemical properties	23
2.3.2	Structure of caveolae and caveolins	24
2.3.3	Function of caveolae	26
2.3.3.1	Caveolae involvement in cellular signaling events	26
2.3.3.2	Caveolae and endocytosis	27
2.3.4	Role of lipid rafts and caveolae in bacterial infections	29
3	**Objectives**	**31**
4	**Results**	**32**
4.1	Microarray analysis of infected AGS vs infected caveolin-1 expressing AGS cells	32
4.2	Confirmation of microarray results by quantitative real-time PCR (qRT-PCR)	34
4.3	Impact of stimulation and inhibition of PKA activity on *N. gonorrhoeae* internalization	37

4.4	*N. gonorrhoeae* activates the PKA pathway of epithelial cells	41
4.5	Recruitment of VASP to infecting *N. gonorrhoeae*	42
4.6	VASP knockdown affects *N. gonorrhoeae* internalization	44
4.7	AC/PKA pathway influences *N. gonorrhoeae* association with caveolin and lysosomal compartments	48
4.8	ErbB2 and Src in *N. gonorrhoeae* infection	52
4.9	*N. gonorrhoeae* triggered actin and VASP dynamics	54
5	**Discussion**	**59**
5.1	AC/PKA pathway in pilus mediated *N. gonorrhoeae* invasion	59
5.2	VASP and *N. gonorrhoeae* infection	62
5.3	Actin recruitment, dynamics and actin-caveolin interplay	63
5.4	Signalling through caveolae/lipid rafts relevant to *N. gonorrhoeae* internalization	65
6	**Conclusion**	**69**
7	**Materials**	**70**
7.1	Bacteria	70
7.1.1	*E. coli*	70
7.1.2	*N. gonorrhoeae*	70
7.2	Cell culture	70
7.3	Cell culture media and supplements	71
7.4	Media for bacterial culture	71
7.5	Plasmid vectors	72
7.6	Oligonucleotides	72
7.7	Antibodies	73
7.8	Buffers and solutions	74
7.9	Chemical reagents	76
7.10	Kits	77
7.11	Appliances and consumable materials	77
7.12	Software	78
8	**Methods**	**79**
8.1	Cell culture methods	79
8.1.1	Passaging of cells	79

8.1.2	Transfection of cells	80
8.1.2.1	Transfection of plasmid DNA	80
8.1.3	Infection of cells with *N. gonorrhoeae*	80
8.1.4	Gentamicin protection assay	81
8.2	Growth and manipulation of bacteria	81
8.2.1	Growth of *N. gonorrhoeae*	81
8.2.2	Growth of *E. coli*	82
8.2.3	Preparation of DNA competent *E. coli*	82
8.2.4	Transformation of *E. coli*	82
8.3	Nucleic acid methods	83
8.3.1	Preparative isolation of plasmid DNA	83
8.3.2	RNA purification from adherent cells	84
8.3.3	Quantitative real-time polymerase chain reaction (qRT-PCR)	84
8.3.4	RNA interference (RNAi)	85
8.4	Protein biochemical methods	86
8.4.1	Discontinuous SDS polyacrylamide gel electrophoresis (SDS-PAGE)	86
8.4.2	Coomassie® stain of protein gels	87
8.4.3	Immuno blot (Western blot)	87
8.4.4	Membrane stripping	88
8.4.5	Protein co-immuno precipitation	88
8.4.6	Purification of PilC	89
8.5	Microscopy	90
8.5.1	Confocal laser scanning microscopy	90
8.5.1.1	Indirect immunofluorescence staining	90
8.5.1.2	Differential indirect immunofluorescence staining of bacteria	90
8.5.1.3	Confocal laser scanning microscopy	91
8.5.2	Life cell imaging microscopy	91
9	**References**	**92**
10	**Index**	**107**
10.1	Figure Index	107
10.2	Abbreviations	109
10.3	Acknowledgements	112

1.1 Abstract

The type VI pili (Tfp) of the Gram-negative bacterium *Neisseria gonorrhoeae* (*Ngo*) are crucial virulence factors for the colonization of its sole natural host, *Homo sapiens*. They do not only mediate the first step in infection by adhering to the host cell surface, but also contribute to other adhesion and invasion processes in concert with other virulence factors like the colony opacity associated (Opa) proteins and lipooligosaccharides (LOS). Furthermore, they are able to promote invasion into some cell lines.

Recently, it was shown that piliated gonococci elicit clustering of the membrane-associated protein caveolin-1 beneath attachment sites, and that downregulation of caveolin by RNA interference (RNAi) increased gonococcal invasion. Complementarily, expression of caveolin-1 in caveolin-deficient epithelial cells blocked internalization. Here, it is demonstrated in a microarray experiment comparing gene regulation of infected epithelial cells with and without caveolin, that the regulatory subunit of the protein kinase A (PKA-RIβ) was highly upregulated in both cell lines. Pharmacological inhibition of PKA resulted in a strong increase of invasion in AGS and ME180 epithelial cells, whereas PKA agonists had the opposite effect. Adenylyl cyclase (AC) and PKA activity were increased during the first two hours of *Ngo* infection, starting around 10 minutes after the addition of bacteria and coinciding with pilus-mediated attachment. The PKA substrate vasodilator stimulated phosphoprotein (VASP) was phosphorylated in response to *Ngo* infection and enriched at the sites of attaching bacteria. Moreover, alteration of PKA activity had strong impact on caveolin-1 recruitment to gonococcal microcolonies. These findings suggest a role for PKA and VASP in the invasion process of *N. gonorrhoeae* by contributing to the assembly of an actin-caveolin network that blocks internalization during localized adherence.

Using life cell imaging microscopy, actin and VASP distribution and dynamics could be monitored, providing insights into the formation of actin clusters beneath gonococcal microcolonies. In addition, actin- and VASP rocketing was also observed in epithelial cells, and infection influenced actin-based comet tails. Whether internalized gonococci, like other microorganisms, have the capacity to induce or hijack actin comet tails to move intracellularly could not ultimately be proven.

1.2 Zusammenfassung

Die Typ IV Pili (Tfp) des Gram-negativen Bacteriums *Neisseria gonorrhoeae* sind wichtige Virulenzfaktoren für die Besiedelung ihres einzigen natürlichen Wirtes *Homo sapiens*. Diese vermitteln den ersten Infektionsschritt durch Adherenz an die Wirtszelloberfläche und tragen, gemeinsam mit anderen Faktoren wie Opa Proteinen und Lipooligosacchariden (LOS), zusätzlich zu anderen Adherenzvorgängen und zu Invasionsprozessen bei. Darüber hinaus verstärken Pili die bakterielle Invasion in bestimmte Zelllinien.

Vor kurzem konnte gezeigt werden, dass pilierte Gonokokken das membranassozierte Protein Caveolin unterhalb der Bindungsstelle auf Epithelzellen akkumulieren und das Ausschalten der Caveolinexpression mittels RNA-Interferenz (RNAi) die Gonokokkeninvasion erhöht. Komplementär hierzu wurde die Invasion durch Caveolinexpression in Zellen, welche normalerweise kein Caveolin synthetisieren, blockiert. In dieser Arbeit konnte bei dem Vergleich von infizierten Zellen mit und ohne Caveolin mit Hilfe eines Microarray-Experimentes eine starke Überexpression der regulatorischen Untereinheit der Proteinkinase A (PKA-RI) in beiden Zelllinien festgestellt werden. Pharmakologische Untersuchungen zeigten einen starken Anstieg der Invasion durch PKA-Inhibierung, eine Behandlung der Zellen mit PKA-Agonisten hingegen resultierte in verminderter Invasion. Die enzymatische Aktivität der Adenylylzyklase (AC) und der PKA war während der ersten beiden Stunden einer Gonokokkeninfektion erhöht. Dieser Effekt trat bereits zehn Minuten nach Bakterienzugabe auf und damit zeitgleich mit dem Beginn der Pilus-vermittelten Adherenz.

Das PKA-Substrat „vasodilator stimulated phosphoprotein" (VASP) wurde durch Neisserieninfektion phosphoryliert und zu den Adherenzstellen der Bakterien rekrutiert. Des Weiteren wurde die Rekrutierung von Caveolin-1 zu bakteriellen Mikrokolonien durch die Beeinflussung der PKA-Aktivität stark verändert. Dies zeigt, dass PKA und VASP in Invasionsprozessen von *N. gonorrhoeae* beteiligt sind, indem diese zum Aufbau eines Caveolin-Actin-Geflechts beitragen, welche die Aufnahme der Bakterien in die Zelle hemmt.

Mittels „life cell imaging" Mikroskopie konnte die Verteilung und Dynamik von Aktin- und VASP-GFP Molekülen beobachtet werden, welche Einblicke in die Entstehung von Aktin-Clustern unterhalb von bakteriellen Mikrokolonien ergaben. Darüber hinaus

Zusammenfassung

wurden dynamische Strukturen, welche als Aktin- bzw. VASP-Kometen bezeichnet werden, in den verwendeten Epithelzellen entdeckt. Diese auf Aktinpolymerisation basierenden Strukturen wurden durch Neisserieninfektion beeinflusst. Ob intrazelluläre Gonokokken diese als Fortbewegungsmittel nutzen oder selbst induzieren, konnte nicht zweifelsfrei geklärt werden.

2 Introduction

2.1 Bacteria – host cell interactions

In the long history of relations between bacteria and their human host, different bacterial species have evolved multiple strategies to survive in their differing biological niches. The different consequences of these strategies, such as host specificity, tropism and pathogenicity, have a basis at molecular level. That is human-microbe relations are determined by interactions of protein, carbohydrate and lipid macromolecules on both sides. Therefore, it is the major goal in infection biology to decipher these molecular interactions occurring between the host cell and the pathogen.

2.1.1 Bacterial adherence to and invasion of target cells

In general, the first step of an infection process is the attachment of the microbe to the host cell. Binding to target cells is either followed by extracellular colonization of the specific tissue or by invasion of cells to establish intracellular accommodation. *Chlamydia* spp for example are obligate intracellular organisms that multiply via a complex life cycle which can only be accomplished inside the cell in a special compartment termed inclusion (AbdelRahman and Belland, 2005). Although adhesion and invasion processes are crucial for *Chlamydia* infection, as they are for the infection of other bacteria, the underlying mechanisms for both events are still poorly understood (Ward and Murray, 1984; Allan and Pearce, 1987).

Pathogens eliciting enteric diseases exhibit elaborate mechanisms to invade host cells. *Salmonella enterica* which causes diarrhoea and typhoid fever, is able to adhere to and invade different cell types such as intestinal epithelial cells, microfold (M) cells and macrophages to overcome the epithelial barrier of the intestine (Haraga *et al.*, 2008). Invasion of epithelial cells is preceded by the delivery of effector proteins by a secretion machinery designated type three secretion system (T3SS) into the cytosol, which leads to the rearrangement of the actin cytoskeleton and, in turn, to the engulfment and uptake of the microbe (Hardt *et al.*, 1998; Unsworth *et al.*, 2004). *Shigella flexneri*, another pathogen that infects the intestinal epithelium (LABREC *et al.*, 1964), lacks, unlike *Salmonella*, adherence factors or flagella. Despite this, *Shigella* efficiently invades M-cells, Macrophages and epithelial cells.

The latter is stimulated by the interaction of the T3SS-effector IpaB to the cellular CD44 receptor, which is present in membrane areas rich in lipid rafts (Skoudy et al., 2000).

Escherichia coli, which normally constitute the gut flora in humans maintains a symbiosis with its host, but can also cause diarrhea. So called enteropathogenic (EPEC) and enterohaemorrhagic *E. coli* (EHEC) colonize the gut mucosa by establishing attaching and effacing (A/E) lesions. These lesions are characterized by loss of microvilli and close association of the bacteria with the enterocyte plasma membrane (Chen and Frankel, 2005). The association of EPEC and EHEC with the host cell is mediated via the adhesin intimin which binds to the translocated intimin receptor (Tir) that is introduced into the host cell membrane via T3SS (Kenny et al., 1997). Actin polymerization, which gives rise to pedestal formation underneath bacteria, is then triggered via the direct binding of the host adaptor protein Nck to Tir or indirectly through the bacterial T3SS injected effector protein TccP/EspF$_U$ (Campellone et al., 2004). Although actin polymerization is not thought to be involved in A/E lesion formation, it plays an as yet unidentified role in colonization (Bai et al., 2008).

The foodborne, Gram-positive pathogen *Listeria monocytogenes*, which causes gastroenteritis, meningitis, encephalitis and mother-to-fetus infections, is able to cross the intestinal barrier by binding to the host cell receptors Met and E-cadherin via the bacterial surface proteins internalin A (InlA) and InlB, respectively, which in turn mediate cytoskeletal rearrangements and internalization of the bacterium (Mengaud et al., 1996; Shen et al., 2000). Furthermore, *Listeria* is able to escape the phagocytic vacuole and propels through the cytoplasm and into neighbouring cells. This kind of locomotion is driven by the polymerization of actin mediated by the bacterial surface protein ActA, which recruits the actin nucleating Arp2/3 complex and the actin polymerization promoting factor VASP (Welch et al., 1997; Niebuhr et al., 1997). A growing number of pathogens, including *Rickettsia* spp, *Shigella* spp, mycobacteria, *Burkholderia pseudomallei* and vaccinia virus show this feature of intracellular movement.

In summary, these examples demonstrate that many bacterial pathogens exploit host cell structures and signaling pathways in order to promote their adherence to or uptake by host cells. For most invasive bacteria, a major consequence of these signaling events is a remodeling of the host cell surface driven by the reorganization of the actin cytoskeleton.

2.2 Neisseria spp.

The genus *Neisseria* belongs to the family of *Neisseriacea* (Murray and Branham, 1948) and contains species that have been isolated from both humans and animals, of which 12 species and serovars are exclusively isolated from humans (Knapp, 1988). *Neisseria* are Gram-negative bacteria of 0.6-1.5 µm in size which occur as diplococci, a property which was responsible for the original assignation of the genus *Neisseria* to the family *Coccacae* until 1948 (Murray and Branham, 1939). Most *Neisseria* species such as *N. lactamica*, *N. sicca* and *N. mucosa* belong to the natural flora of the human oro- and nasopharynx and are considered commensal bacteria, although they can occasionally act as opportunistic pathogens (Johnson, 1983). In addition to the commensal species, two members of the genus *Neisseria* are considered pathogens: *N. gonorrhoeae* (*Ngo*), which is always pathogenic and the etiological agent of the sexual transmitted disease gonorrhea, and *N. meningitides* (*Nmg*), which can cause meningitis. Both pathogenic *Neisseria* species are neither able to colonize other animals, nor do they survive in the environment, which limits them to humans.

2.2.1 Meningococcal pathogenesis

N. meningitides, the meningococcus, colonizes the nasopharynx of humans and exhibits a commensal life style in the vast majority of carriers. The prevalence in healthy individuals is up to 30% in non-epidemic geographic areas (Caugant *et al.*, 1994; Claus *et al.*, 2005). In some cases, however, *Nmg* can invade the mucosal epithelium and enter the bloodstream, causing septicaemia, and are able to overcome the blood-brain barrier, which results in fulminant meningitis (Booy and Kroll, 1998; Nassif, 1999). The triggers for invasive meningococcal infections are not fully understood, but they are facilitated by viral co-infection (Griffiths *et al.*, 2007), dry air or other noxa (smoking, alcohol etc.). Clones collected for epidemiologic studies appear less pathogenic than those isolated from patients with acute meningococcal infection, which argues for the hypothesis that only certain *Nmg* strains exhibit an invasive capacity (Yazdankhah *et al.*, 2004). In support of this, a recent study correlates invasiveness of meningococci with the presence of a bacteriophage in a subclass of highly invasive strains (Bille et al., 2005).

In underdeveloped regions (e.g. the "meningitis belt" in Southern Sahara), epidemics of Nmg infections are a major cause of morbidity and mortality, causing 10,000 or more deaths in a single outbreak (Hart and Cuevas, 1997). More than 90% of all meningococcal infections are caused by the serotypes A, B, C and Y, and vaccines are available for the serotypes A, C, Y and W135. In Germany, the majority of invasive Nmg infections are caused by the serotypes B (67%) and C (18%). Out of 555 cases in 2006, which equals an incidence rate of 0.67 cases per 100,000 inhabitants, 53 did not survive the infection, which yields a mortality rate of about 10% in Germany (Vogel et al., 2007).

2.2.2 Gonococcal pathogenesis and clinical manifestations

N. gonorrhoeae is closely related to N. meningitides and typically colonizes the mucosal epithelia of the male urethra and the female uterine cervix, but infections can also occur on the rectum, the throat and the conjunctiva of the eye. The latter can easily happen to newborns if the mother carries the pathogen. Hence, as a prophylaxis, the eyes of newborns are treated with antimicrobial agents (e.g. silver nitrate) to avoid neonatal ophthalmia infection during birth.

Ngo transmission generally occurs through direct sexual contact, but indirect modes of transmission have been reported (Kleist and Moi, 1993). Attachment to mucosal epithelium is followed within 24-48 hours by penetration of the organism between and through epithelial cells to the submucosal tissues. This is followed by massive infiltration of the infected epithelial tissue by neutrophils, resulting in the development of submucosal microabscesses and exudation of pus. Gram-stained urethral exudates reveal a large number of gonococci associated intra- and extracellularly with a few neutrophils. It is assumed that some gonococci are able to evade killing mechanisms and continue to multiply intracellularly (Casey et al., 1986). Some immune evasion strategies like the secretion of an IgA protease, antigenic- and phase variation of pili and colony opacity-associated (Opa) outer membrane proteins are already well described. Due to the antigenic variation of the pathogen, a vaccine against N. gonorrhoeae is not available and an endured infection does not confer immunity.

About 60 million gonococcal infections occur each year worldwide, despite effective antibiotic therapies. This phenomenon can be explained by the fact that up to 15% of infected men and 80% of infected women remain without symptoms (Handsfield, 1990). Sexually mature women and newborns, under the influence of estrogen, are often protected from vaginitis due to cornification of the vaginal epithelium because *Ngo* primarily infects noncornified epithelium. Infected women who become symptomatic do so within 10 days, and the dominant symptoms are those of cervicitis and/or urethritis, including increased vaginal discharge, dysuria, intermenstrual bleeding and menorrhagie (McCormack *et al.*, 1977; Barlow and Phillips, 1978). In 10-20% of the cases, the infection, if not treated, ascends the endometrium and the fallopian tubes, resulting in pelvic inflammatory disease (PID), which is manifested by various combinations of endometritis, salpingitis (infection of fallopian tubes), tubo-ovarian abscess and pelvic peritonitis (Holmes *et al.*, 1980). The consequence of these incidences are serious sequelae, such as ectopic pregnancies and infertility due to fallopian tube obstruction (Sweet, 1987). In men, a gonococcal infection is evident by acute urethritis with urethral discharge of pus and dysuria as the major symptoms and acute epididymitis as the most common complication. Disseminated gonococcal infection (DGI) is rare (0.5-3% of infected patients), but can lead to septic arthritis and dermatitis as the predominant manifestations, or even to meningitis (Sayeed *et al.*, 1972) and osteomyelitis (Masi and Eisenstein, 1981).

2.2.3 *N. gonorrhoeae* adhesion to and invasion of epithelial cells

In order to adhere to and invade host cells, *N. gonorrhoeae* has developed several different factors, and the list of reported factors involved in adhesion and invasion processes is growing continuously (Paruchuri *et al.*, 1990; Capecchi *et al.*, 2005; Du and Arvidson, 2006; Serino *et al.*, 2007). In this paragraph, the major and thus far best characterized factors will be described (Fig. 2-1).

2.2.3.1 Type IV pili

Type IV pili (Tfp) are filamentous polymers of the pilE protein 60-80 Å in diameter and up to several micrometers in length. These structures are found in many different Gram-negative bacteria such as *Vibrio cholerae* (Taylor *et al.*, 1987), *Salmonella enterica* (Zhang *et al.*, 2000) and *Escherichia coli* (Giron *et al.*, 1991). Tfp not only mediate the first adhesion step of Ngo to epithelial cells, but also account for bacterial aggregation (i.e. microcolony formation), twitching motility and natural DNA competence.

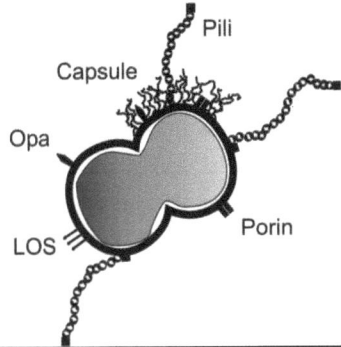

Figure 2-1: Neisserial factors involved in adhesion and invasion
Displayed structures all contribute to the adhesive and invasive properties exhibited by gonococci. The polysialic acid capsule, however, is only expressed by meningococci. It confers serum resistance and acts as an anti-adhesin, since it prevents outer membrane factors like Opa proteins and porin from recognizing their receptors.

Pili, like Opa proteins and lipooligosacharide (LOS; see below), undergo phase- and antigenic variation (Meyer et al., 1988; Meyer and van Putten, 1989), a mechanism which enables gonococci to circumvent host immune response by structurally altering the parts of the major pilus subunit pilE which are exposed to the pilus surface and thus accessible to the host immune system. The underlying mechanism of this phenomenon is mainly a unidirectional transfer of a portion of one of several silent pilE homologous genes (pilS) via DNA-recombination to the pilE expression locus, although pili variation can also result from transformation by exogenous DNA (Gibbs et al., 1989). Antigentic variation, which occurs at a rate of 10^{-4} per cell division, can also result in a pilE variant defective in fiber formation, leading to a non-piliated phenotype.

PilE or pilin monomers, which result from the cleavage of prepilin by the peptidase PilD, are assembled into a pilus fiber in the inner membrane of the bacterium and secreted through a pore in the outer membrane consisting of a dodecamer of the minor pilus subunit PilQ (Wolfgang et al., 2000). This process is thought to be driven by the PilC protein by regulating PilT (Morand et al., 2004), a pilus subunit that is hypothesized to mediate pilus disassembly by hydrolyzing ATP, causing retraction of the pilus (Whitchurch et al., 1991). PilT knockout mutants are hyperpiliated, do not twitch and microcolonies formed by this mutant appear aberrant in shape (Higashi et al., 2007). Retraction of a single pilus was shown to generate a considerable force of 50-100 piconewton (Merz et al., 2000), and recent findings provide evidence that bundles of up to 8-10 Tfp have the capacity to exert forces in the nanonewton range (Biais et al., 2008), which makes PilT the strongest biological motor known to date.

In addition to its role in PilT regulation, PilC was shown earlier to be the tip-located adhesin of the pilus (Rudel et al., 1995), but the cellular receptor for PilC has not been identified yet. Pili are also able to agglutinate erythrocytes in an PilC independent fashion (Scheuerpflug et al., 1999) and different pilE variants, which result fom antigentic variation events, exhibit dramatic differences in pilus mediated attachment (Nassif et al., 1993; Jonsson et al., 1994), arguing that PilE also has adhesive properties. Another minor pilus subunit, PilF, a putative ATPase of the Gsp/PulE family, is essential for Ngo pilus assembly (Freitag et al., 1995).

In an overlay assay experiment using ME180 cell lysate immobilized on a blot membrane that was incubated with a pili preparation, the surface protein membrane cofactor protein (MCP) or CD46, implicated in the regulation of complement activation, was detected and therefore proposed as pilus receptor (Kallstrom et al., 1997). However, anti-CD46 antibodies do not block binding to primary cervical and male urethral epithelial cells (Edwards et al., 2002; Edwards and Apicella, 2005) and pilus-mediated attachment independent of CD46 can be observed (Kirchner et al., 2005). Recently, the inserted (I)-domain of β_2-, α_1- and α_2-integrins has been proposed as a pilus receptor (Edwards and Apicella, 2005).

Purified pili and infection with pili-expressing Ngo both elicit a calcium transient in treated epithelial cells occurring minutes after treatment (Kallstrom et al., 1998; Ayala et al., 2005). Calcium release is dependent on intracellular calcium stores and triggered by a precedent calcium influx from the extracellular medium evoked by porin (see below). The first, porin-dependent calcium transient is believed to trigger endosome exocytosis and LAMP1 redistribution to the plasma membrane (Ayala et al., 2002) while lysosome exocytosis seems to be the result of the second, pilus-dependent calcium release (Ayala et al., 2001).

Upon attachment to epithelial cells, gonococci trigger a pili dependent, Opa independent formation of cortical plaques, beginning within minutes after attachment and culminating 4-6 hours post infection. These plaques are characterized by the rearrangement of the cortical cytoskeleton and the plasma membrane, including the accumulation of actin, ezrin, and transmembrane glycoproteins like the epidermal growth factor receptor (EGFR or ErbB1), CD46 and the cell adhesion molecule ICAM-1 (Merz and So, 1997; Merz et al., 1999). Moreover, a concentration of tyrosine phosphorylated proteins is observed underneath microcolonies. This appears to be a recruitment of already phosphorylated proteins rather than a *de novo* phosphorylation (Merz and So, 1997). The receptor tyrosine kinase ErbB2 is also found beneath *Nmg* microcolonies and becomes phosphorylated during meningococcal localized adherence. Inhibition of ErbB2 phosphorylation inhibits *Nmg* invasion of endothelial cells (Hoffmann et al., 2001).

8-16 hours after pilus mediated adherence, piliation is lost, thereby allowing a more intimate attachment of the bacterium via Opa proteins, porin and LOS, all of which are involved in uptake processes. This step is accompanied by the transition of localized adherence of *Neisseria* in microcolonies (10-100 bacteria) to a more diffuse adherence to epithelial cells (Pujol et al., 1997). The pilT protein in *N. meningitides* has been demonstrated to be required for this transition because an *N. meningitides* PilT deletion mutant was unable to change from localized to diffuse adherence and does not show intimate attachment with the epithelial cell membrane (Pujol et al., 1999). However, this mutant also revealed higher adhesion and invasion compared to the wild type strain. Whether the loss of piliation is a prerequisite for invasion seems to be cell type specific since pili enhance the invasion of T84 polarized epithelial cells (Merz et al., 1996) but do not promote entry into Chang cells. In an Opa positive background, pili even remarkably inhibit invasion of Chang cells (Makino et al., 1991) and a non-piliated state of bacteria facilitates multiplication within and exit from T84 polarized epithelial cells (Criss and Seifert, 2006). The majority of bacteria isolated from individuals with acute gonorrhea are piliated, suggesting a selection for pilus expression *in vivo* (Kellogg, Jr. et al., 1963; Swanson et al., 1987; Seifert et al., 1994). However, it is hypothesized that this selection is not due to the better survival of piliated bacteria within invaded cells, but rather stems from a better extracellular survival and greater likelihood of recovery of piliated gonococci compared to non-piliated gonococci from mucosal exudates (Criss and Seifert, 2006).

2.2.3.2 Opa proteins

Subsequent to the binding via pili, the colony opacity-associated (Opa) proteins, a family of phase variable outer membrane proteins, confer intimate attachment to and trigger invasion of target cells (King and Swanson, 1978; Heckels, 1981). A single gonococcal strain can harbor up to 12 different *opa* genes, which show a rapid on/off-switching in their expression status due to phase variation. This variation is the result of a modulation of the repeat number of pentameric sequence repeats within the coding region by slipped-strand mispairing. Since the repeat number is crucial for the correct reading frame, expression is stochastically impaired on the translation level, generating heterogeneous gonococcal populations with individual bacteria featured as distinct cellular tropisms (Makino et al., 1991; Kupsch et al., 1993).

Only one Opa protein out of the 12 present in the gonococcal MS11 strain, Opa_{30} (in general Opa_{HS}), recognizes heperansulphate proteoglycan (HSPG) on the surface of human cells. Furthermore, Opa_{30} and other Opa_{HS} are able to bind extracellular matrix (ECM) proteins such as vitronectin and fibronectin and are thereby able to indirectly bind to integrins (Gomez-Duarte et al., 1997; van Putten et al., 1998b). Because both, HSPG and integrins are mainly expressed at the basolateral side of polarized epithelial cells, Opa_{30} is thought to mediate adhesion and invasion after bacteria have overcome the mucosal epithelial barrier by transcytosis. Inhibition of the lipid modifying enzymes phosphatidylcholine-dependent phospholipase C (PC-PLC) and acid sphingomyelinase (ASM) by pharmacological and genetic means blocks initiated invasion of Chang conjunctiva cells. Since the Opa-induced activity of PC-PLC and ASM leads to the release of diacylglycerol and ceramide, respectively, it may be argued that these second messengers contribute to cytoskeletal rearrangements during gonococcal invasion (Grassme et al., 1997).

Beside the Opa_{HS} proteins engaging HSPGs, the vast majority of these outer membrane proteins, designated Opa_{CEA}, mediate binding to the carcinoembryonic (CEA) related cell adhesion molecule (CEACAM) family that harbour the CD66 epitope. The human CEACAM family, which is involved in neutrophil binding to epithelial cells (Kuijpers et al., 1992), comprises seven members (CEACAM1, CEACAM3-8), of which CEACAM4, CEACAM7 and CEACAM8 are not recognized by any Opa protein characterized thus far. Although CEACAMs are mainly restricted to the hematopoetic cell lineage and are not expressed by many epithelial cell lines, they are upregulated in epithelial cells derived from the female genital tract (Muenzner et al., 2002) and in primary endothelial cells (Muenzner et al., 2001) upon *N. gonorrhoeae* infection. Moreover, in contrast to HSPGs, CEACAM receptors are apically expressed in polarized epithelial cells and are therefore accessible to bacteria from the lumenal side, enabling gonococci to efficiently bind and transcytose polarized T84 epithelial cells in an Opa_{30} dependent manner (Wang et al., 1998). CEACAMs are also exploited by other pathogens and, interestingly, commensal Neisseria were also found to express pathogen-like Opa adhesins that are able to interact with CEACAM1 (Toleman et al., 2001). Together with the findings that Opa proteins are not needed for initial colonization of the host, are not required for adherence or invasion of primary cervical epithelial cells (Edwards et al., 2002) and are not essential for the process of invasion (Swanson et al., 2001), these observations suggest that, in addition to Opa proteins, other virulence factors (e.g. Tfp) are crucial to enable gonococcal internalization into epithelial cells.

2.2.3.3 Lipooligosaccharides (LOS)

LOS molecules, which are also referred to as LPS (lipopolysaccharides), consist of three oligosaccharide chains associated to a lipid A core (Yamasaki et al., 1991). The oligosaccharide moiety of LOS, which is similar in sequence and linkage to human oligosaccharides, is composed of a highly conserved core structure and a terminal oligosaccharide which undergoes rapid phase variation, resulting in the expression of several terminal LOS structures on the outer membrane of N. gonorrhoeae at any given time (Burch et al., 1997). The asialoglycoprotein receptor (ASPG-R) is the binding site of LOS containing lacto-N-neotetraose, and ASPG-R expression in hepatoma cells is upregulated by gonococci (Porat et al., 1995). Furthermore, LOS containing lacto-N-neotetraose facilitates the invasion of Ngo into ME180 (Song et al., 2000) and primary urethral epithelial cells (Harvey et al., 2001), and receptor mediated endocytosis with involvement of clathrin-coated pits is reported to be the underlying mechanism, at least in the latter case. It is important to note that Ngo invasion of ME180 cells is pili-dependent, and non-piliated phase variants show strongly reduced adhesion and almost no invasion (Song et al., 2000), underscoring the relevance of pilus expression in gonococcal infection.

Both, gonococcal and meningococcal LOS can be modified by the addition of terminal sialic acid moieties (Smith et al., 1995; Preston et al., 1996). On one hand, this modification significantly inhibits LOS- and Opa/Opc-mediated invasion and adhesion, respectively (van Putten, 1993; Virji et al., 1995; Harvey et al., 2001); on the other hand, it confers resistance to complement and ingestion by professional phagocytes (Vogel and Frosch, 1999). Gonococci are not able to synthesize sialic acid, and the sialyl donor for *Ngo* LOS is host derived cytidinemonophosphate-*N*-acetylneuraminic acid (CMP-NANA) (Smith *et al.*, 1992). The *Nmg* capsule, which is composed of long polysialic acid chains, is, like sialic acid modified LOS, antiphagocytic and antibactericidal since it increases the bacterium's negative surface charge density (Vogel and Frosch, 1999). The capsule is thus universally expressed by strains isolated from the blood and cerebrospinal fluid of patients with meningococcaemia.

2.2.3.4   Porin

Neisserial Porins belong to the Gram-negative porin superfamily (Jeanteur *et al.*, 1991) and are the most abundant outer membrane proteins in pathogenic *Neisseria*. There are two different porin genes: PorA and PorB in meningococci and PorB1A and PorB1B in gonococci. Porins from *Nmg* and *Ngo* share 60-70% amino acid homology and moderate antigenic variation, which is the basis of the neisserial serotyping system (Frasch *et al.*, 1985).

Porins form anion-selective anion channels which are essential for neisserial viability, consistent of three single polypeptides each containing a high proportion of β-pleated strands. These confer a predicted β-barrel structure motif to the native trimeric molecule (Minetti *et al.*, 1997). They are able to interact with planar lipid bilayers and eukaryotic target cell membranes by inserting into them (Weel and van Putten, 1991), causing a calcium influx into *Ngo* infected cells from the extracellular milieu (Muller *et al.*, 1999) independently of pili. The translocation mechanism of porins into host cells is poorly understood to date.

Gonococcal PorB1A is associated with increased bacterial invasiveness, and strains harboring PorB1A, which are responsible for disseminated disease, are more resistant to killing by normal human serum and invade cells *in vitro* to a greater extent than gonococci bearing the PorB1B protein (van Putten *et al.*, 1998a). Recent findings suggest that the glycoprotein Gp96 serves as a binding site for gonococci expressing PorB1A and that the Gp96-associated scavenger receptor is responsible for PorB1A-dependent gonococcal uptake into host cells (Rechner *et al.*, 2007). However, porin-dependent *Ngo* adherence to and invasion of eukaryotic cells has only been demonstrated in the absence of inorganic phosphate in the infection medium, which might be explained by differences in the modulation of porin function (Rudel et al., 1996). In the case of Nmg, porins have not been associated with bacterial entry into host cells, although porins show the capacity to induce actin polymerization, which argues for a possible implication in host cell actin reorganization (Wen et al., 2000).

Controversial data has been collected about the role of porin in apoptosis. Whereas some findings suggest that gonococcal porin, upon a calcium influx into host cells and its translocation to the host mitochondria, triggers cytochrome c release and apoptosis (Muller et al., 2000), other observations reveal a protective effect of meningococcal porin against the induction of apoptosis by staurosporine (Massari et al., 2000). Besides the different sources of porin, this contradiction can in part be explained by the different experimental settings, since fetal calf serum (FCS) was present in the cell culture medium of one experiment but not in the other, and FCS was found to be an efficient inhibitor of the pro-apoptotic effect of porin (Muller et al., 2000).

2.2.3.5 Models for N. gonorrhoeae invasion

Thus far, there is no detailed model of the molecular invasion mechanisms involved in the in vivo N. gonorrhoeae infections. This is due to the complex interplay of the above mentioned adhesion and invasion factors, which are mostly studied in strains that lack one or the other component to circumvent and dissect the functional overlapping.

However, some uptake mechanisms have been associated with certain invasins or even invasin combinations. As described above, LOS-dependent uptake of Ngo is thought to occur through receptor-mediated endocytosis of clathrin-coated pits (Harvey et al., 2001). Nonopsonic phagocytosis in professional phagocytes is thought to be triggered by Opa-CD66 (CEACAM) engagement and involves, amongst others, the activation of Src-like tyrosine kinases and Rac1 (Hauck et al., 1998). CEACAM1 transfected HeLa cells exhibit cellular protrusions tightly enveloping bacteria reminiscent of engulfment events characteristic for phagocytes in a Rac and Cdc42 independent fashion (Billker et al., 2002). Although the small GTPases Rac and Cdc42 are activated upon Opa-CEACAM3 interaction and CEACAM3-phosphorylation by Src kinases, this is not the case for CEACAM1. Furthermore, bacteria-engulfing actin structures are sensitive to dominant interfering GTPase constructs and the actin-disrupting drug cytochalasin D in CEACAM3, but not in

Introduction

CEACAM1 expressing epithelial cells (Billker et al., 2002). Since CEACAM1, but not CEACAM3 is expressed by epithelial cells, it is likely that different mechanisms account for the Opa-dependent invasion of professional phagocytes and the epithelium.

Another invasion model involving Src kinase was established with piliated, Opa negative meningococci. This model postulates two distinct signaling pathways that both mediate the microvilli-like protrusion formation which engulf the bacteria and promote their uptake into endothelial cells (Fig. 2-2).

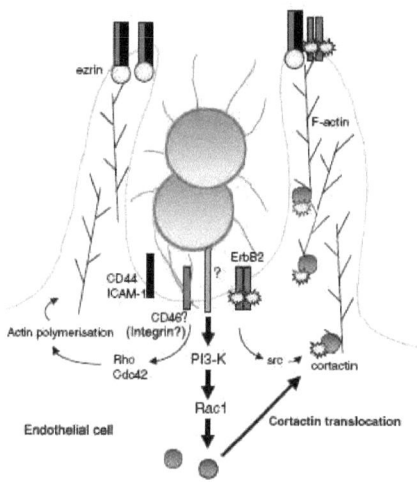

Figure 2-2: Model for *N. meningitides* invasion
Meningococci attach to endothelial cells via pilus binding to a cellular receptor. Ezrin, which links the cellular membrane to the actin cytoskeleton, and several transmembrane proteins (ICAM-1, ErbB2, CD46) are recruited to the bacterial adhesion site. In a second step, bacteria trigger Rho and Cdc42 GTPases activation, which in turn, induce polymerization of cortical actin. This leads to the formation of membrane projections which engulf single bacteria, resulting in its internalization. Clustering of ErbB2 triggers its autophosphorylation and subsequent activation of cortactin, which influences actin filament architecture and thus bacterial uptake. Cortactin recruitment is mediated via LOS-dependent PI3-kinase and subsequent Rac activation (Lambotin et al., 2005; modified).

Both pathways depend on pilus-induced clustering of different cellular factors like ErbB2, CD44/ICAM-1 and ezrin. In one pathway, a Rho and Cdc42 dependent actin polymerization orchestrated by ezrin results in membrane protrusions (Eugene et al., 2002). In the other, two concomitant signaling events are integrated to finally result in a contribution to protrusion formation. Bacterial LOS triggers a PI3-kinase and Rac dependent recruitment of cortactin to the site of bacterial adhesion (Lambotin et al., 2005). Cortactin is then phosphorylated by Src that is activated by the receptor tyrosine kinase ErbB2 upon pilus-induced homodimerization of the receptor (Hoffmann et al., 2001). Phosphorylated cortactin then promotes actin polymerization and influences the actin architecture of membrane protrusions (Lambotin et al., 2005). Whether this model is also valid for N. gonorrhoeae remains to be determined. Apparently, several differences in the uptake mechanism between the two pathogens exist. For example, the receptor for gonococcal LOS, ASPG-R (Harvey et al., 2001), is not expressed by endothelial cells which were applied for meningococcal studies (Lambotin et al., 2005). Furthermore, EGF receptor recruitment is observed in cortical plaques beneath bacterial attachment sites in N. gonorrhoeae (Merz et al., 1999), but not in N. meningitides infection (Hoffmann et al., 2001).

2.3 Caveolae and lipid rafts

2.3.1 Biochemical properties

Biological membranes were long thought to be best described by the "fluid mosaic" model, which envisions membranes as a sea of homogenously distributed lipids in which proteins can float and diffuse freely (Singer and Nicolson, 1972). This view has been refined by the model introduced by Simons and others, which claims that membrane proteins are arranged much more heterogeneously and can, in part, be found in so called lipid rafts, specialized membrane microdomains rich in cholesterol, glycosphingolipids and glycosylphosphatydylinositol (GPI) anchored proteins (Simons and Ikonen, 1997). Because of the higher content of saturated fatty acyl chains in sphingo- and phospholipids, which allow a tighter packing than cis-unsaturated phospholipids, lipid rafts exhibit a higher melting temperature than the bulk membrane fraction (Schroeder et al., 1994). In association with the high cholesterol concentration, lipid rafts are resistant to extraction with nonionic detergents such as Triton X-100 at 4°C (Schroeder et al., 1998; Brown and London, 1998) and, due to the relatively higher lipid to protein ratio, have a low density. These properties enable their purification as the light buoyant density fraction in sucrose gradient centrifugation of membrane preparations (Brown and Rose, 1992).

Rafts are thought to be small, 40-50 nm in diameter structures which are highly dynamic and contain a limited number of proteins, approximately less than ten. However, crosslinking agents (e.g. ligands, antibodies) are able to drive the association of rafts, resulting in raft clusters which bring different raft associated proteins together.

Whereas lipid rafts are not detectable by microscopical means, caveolae have already been identified in the 1950s. Although the overall composition of lipid rafts and caveolae is not identical, the main difference between the two is a protein exclusively found in caveolae, termed caveolin. In addition to caveolin, several proteins have been shown to preferentially localize to either caveolae or lipid rafts, respectively (Liu *et al.*, 1997).

2.3.2 Structure of caveolae and caveolins

In contrast to the flat lipid rafts which are indistinguishable from other membrane parts, caveolae are 50-100 nm flask-shaped invaginations of the plasma membrane, which occupy 20% of the plasma membrane of adipocytes where they are most abundant. Simultaneously identified in capillary endothelial cells and epithelial cells from the mouse gall bladder (Palade, 1953; YAMADA, 1955), caveolae are present to some degree in most differentiated cell types. In particular, they are well described in adipocytes, endothelial cells, type I pneumocytes of the lung, and striated and smooth muscle cells. The cytoplsamic surface of caveolae was found to be coated by numerous striated filaments, and an antibody recognizing a 22 kDa protein was used to stain this striation in immunoelectron microscopy experiments, which was therefore termed caveolin (Rothberg *et al.*, 1992).

Three members of the caveolin gene family have been identified to date. Caveolin-1 was originally identified as a tyrosine-phosphorylated 22 kDa substrate of v-Src in transformed chicken embryonic fibroblasts (Glenney, Jr., 1989). Subsequent investigations revealed that it occurs in two isoforms, termed α and β, which, respectively, comprise amino acid residues 1-178 and 32-178, with the latter resulting in a protein ~3 kDa smaller in size (Scherer *et al.*, 1995). Both isoforms are able to form caveolae, and the exact functional significance of these distinct isoforms remains unclear. However, the caveolin-1α isoform can drive the formation of caveolae more efficiently (Fujimoto *et al.*, 2000). Caveolin-2, which occurs in three isoforms (α-γ), is only expressed in the presence of caveolin-1 that prevents its degradation by the proteasome and allows caveolin-2 transport from the Golgi apparatus to the plasma membrane (Mora *et al.*, 1999). The expression of caveolin-3 is restricted to muscular tissues, where it is the only caveolin family member present (Tang et al., 1996). As an exception, this is not the case for smooth muscles, which express all caveolin genes.

Caveolins are membrane proteins attached to the cytoplasmic leaflet of the plasma membrane. In the case of caveolin-1, the NH_2- and COOH-terminus, which are tyrosine phosphorylated (Li et al., 1996) and palmitoylated (Dietzen et al., 1995), respectively, are separated from each other by a short membrane insertion of 32 hydrophobic amino acids (residues 102-134). This sequence forms a unique hairpin loop configuration which prevents caveolin from completely spanning the plasma membrane (Monier et al., 1995). In addition, membrane interaction is mediated by two domains flanking the hydrophobic region, the NH_2-terminal membrane attachment domain (N-MAD) and the COOH-terminal membrane attachment domain (C-MAD).

Besides its membrane association properties, N-MAD is able to bind proteins that possess one of the two related but distinct caveolin binding motifs $\phi xxxx\phi xx\phi$ and $\phi x\phi xxxx\phi$, where ϕ is an aromatic amino acid and x is any amino acid (Couet et al., 1997a). Also termed caveolin scaffolding domain (CSD), N-MAD acts as an anchor capable of restricting proteins to caveolae compartments as well as a regulatory element either inhibiting or enhancing a protein's signaling activity.

Caveolin-1 and, similar, caveolin-3 form high molecular mass oligomers of ~400 kDa, and this homo-oligomerization is mediated by a domain mapping to residues 61-101 which is referred to as the oligomerization domain (Sargiacomo et al., 1995). It is thought that the capability of forming homo-oligomeric complexes contributes to the formation of caveolae since caveolin-2 does not form these complexes (Scherer et al., 1996) and is not able to form caveolae by itself. However, other factors such as cholesterol are also involved in membrane invagination as the depletion of cholesterol with methyl-β-cyclodextrin leads to the flattening of caveolae (Thorn et al., 2003) and neurons which lack caveolae (Lang et al., 1998) express all three caveolin family members (Ikezu et al., 1998).

2.3.3 Function of caveolae

2.3.3.1 Caveolae involvement in cellular signaling events

Caveolae function has been broadly studied during the last decade, and its role in modulating cell signal transduction has been one of the major foci. As a starting point, Lisanti and colleagues showed that biochemically purified caveolae microdomains contain signaling molecules such as Src-like tyrosine kinases and heterotrimeric G proteins (Sargiacomo et al., 1993).

Signaling of receptor tyrosine kinases and, in particular, the epidermal growth factor (EGF) receptor in conjunction with caveolae/lipid rafts were under intense investigation in the past. Although the EGF receptor, in contrast to most other receptors examined thus far, does not localize to caveolae, it partitions with the lipid raft compartment, and EGF binding as well as tyrosine kinase activity were shown to be negatively regulated by cholesterol. However, the EGF receptor, upon stimulation with EGF, could be immunoprecipitated with caveolin (Matveev and Smart, 2002) and bears a recognition motif for the caveolin scaffolding domain. These findings argue for the possibilitiy that the receptor can, as an alternative to clathrin-coated pits, be internalized by caveolae. Recently, it was reported that EGF induces caveolin-1 phsosphorylation at tyrosine 14 and thereby stimulates caveolae formation and dynamics (Orlichenko et al., 2006).

Caveolin is thought to be both, a negative regulator of the vast majority of signaling proteins with which it interacts as well as a tumor suppressor. Additionally, considering its widespread tissue distribution, it is remarkable, if not surprising that caveolin-1 null mice do not form spontaneous tumors and are viable and fertile. However, cultured primary mouse embryonic fibroblasts (MEFs) derived from caveolin-1 knockout mice exhibit a marked increase in growth rate (Razani et al., 2002). Furthermore, if crossed with MMTV-PyMT mice, a tumor-prone transgenic model of breast cancer, caveolin-1 null mice develop multifocal dysplastic lesions at a much earlier age than MMTV-PyMT mice alone (Williams et al., 2003). It thus appears that the absence of caveolin alone is insufficient to induce cell transformation in vivo, but leads to this in combination with a transforming agent.

The PI3-kinase independent Insulin signaling pathway seems to involve caveolae. Receptor stimulation by insulin binding triggers receptor autophosohorylation, the first step of a signaling cascade which finally leads to the activation of a raft-localized Rho family GTPase, TC10, and, subsequently, translocation of the GLUT4 glucose transporter to the plasma membrane. If raft integrity or localization of TC10 to rafts is impaired, insulin fails to stimulate glucose uptake (Watson et al., 2001). There is an ongoing debate on whether the insulin receptor partitions with lipid rafts or with caveolae. Contradictory results are found in the literature for many different proteins allocated to cholesterol rich membrane domains. Differences in the preparation procedure are most likely the cause for this problem since many cell fractionation protocols do not separate caveolae from lipid rafts (Pike, 2005). At least, caveolin does play a critical role in insulin signaling because caveolin-1 null mice show markedly decreased glucose uptake compared to wild-type control animals (Cohen et al., 2003).

Another pathway whose components are associated with caveolae is that of the adenylyl cyclase/PKA. Both, adenylyl cyclase and protein kinase A (PKA) bind and are inhibited by caveolin or caveolin-derived peptides (Toya et al., 1998; Razani et al., 1999), and the upstream factor β-adrenergic receptor was also shown to colocalize with caveolae (Schwencke et al., 1999).

2.3.3.2   *Caveolae and endocytosis*

Beside the well characterized uptake of membrane and specific plasma membrane proteins by clathrin coated pits, which is already described in considerable molecular detail (Conner and Schmid, 2003), lipid rafts/caveolae are, among other alternatives, also thought to be involved in endocytic processes. This assumption stems from three main observations. First, nearly all molecules that are known to be internalized independently of clathrin are found in lipid rafts, and typical representatives for clathrin-dependent internalization are excluded from these membrane micro domains (Nichols and Lippincott-Schwartz, 2001). Second, cholesterol depletion blocks uptake of many molecules reported to be internalized by clathrin independent mechanisms. Third, treatment with the phosphatase inhibitor okadaic acid causes budding of caveolae (Parton et al., 1994) and dynamin, a GTPase involved in budding of clathrin-coated pits is also found at the necks of caveolae (Oh et al., 1998).

Although caveolar budding is thought to occur at high rate in endothelial cells, experiments with different epithelial standard cell culture models showed that caveolin is largely immobile at the plasma membrane (Thomsen et al., 2002). The actin binding protein filamin also binds to caveolin (Stahlhut and van Deurs, 2000), and this link suggests that it is responsible for the immobility of plasma membrane caveolae. The uptake of Gp60, a receptor for albumin localized to caveolae in endothelial cells, is inhibited by the overexpression of caveolin-1 (Minshall et al., 2000). Moreover, decreased expression of caveolin-1 in transformed NIH-3T3 cells coincides with increased clathrin-independent uptake of autocrine motility factor, and this effect is reversed when caveolin-1 expression is increased by adenovirus transduction (Le et al., 2002). These observations led to the development of two models which do not necessarily exclude each other. One postulates the existence of transient caveolar invaginations during budding into the cell which do not contain caveolin, and the expression of caveolin stabilizes these invaginations, thereby preventing them from budding (Nabi and Le, 2003). The second model views caveolae endocytosis as a highly regulated process, suggesting two distinct pools of caveolae, a static one and one involved in budding events (Fig. 2-3). Certain stimuli then shift stable caveolar structures to endocytically competent ones capable of pinching off from the plasma membrane. Support for both models comes from experiments on the endocytic route of simian virus 40 (SV40). The virus, upon attachment to MHC class I molecules on the cell surface (Breau et al., 1992; Stang et al., 1997; Pelkmans and Zerial, 2005), triggers an increase in caveolae turnover (Pelkmans and Zerial, 2005) and enters the cell via caveolae (Anderson et al., 1996). However, SV40 also enters cells devoid of caveolin-1 (Damm et al., 2005), and siRNA-mediated silencing of caveolin-1 reduces infectivity only by about 50% (Pelkmans et al., 2004), which suggests that the virus might use both endocytic routes outlined above.

Endocytosed caveolin/lipid raft vesicles dock onto and fuse with at least two endocytic compartments, caveosomes and early endosomes (Pelkmans et al., 2004). Since they do not disassemble the caveolar coat, vesicles maintain their identity during transient interaction with endosomes. Expression of a dominant-active mutant of the small GTPase Rab5, which regulates membrane traffic towards early endosomes, leads to a more permanent association of vesicles with endosomes (Pelkmans et al., 2004). Furthermore, a dominant active mutant of Arf1, another small GTPase which is responsible for membrane traffic from and to the ER-Golgi intermediate (ERGIC) and the cis-Golgi compartment, results in the sequestration of stable caveolar domains on an enlarged golgi complex. Before being delivered to the endoplasmatic reticulum (ER), SV40 accumulates in what is now referred to as caveosome, an endocytic vesicle with a neutral pH containing caveolin-1 but devoid of markers of clathrin-mediated endocytosis (Pelkmans et al., 2001). Through this itinerary, the virus bypasses the lysosomal compartment and circumvents its degradation (Pelkmans et al., 2001; Norkin et al., 2002).

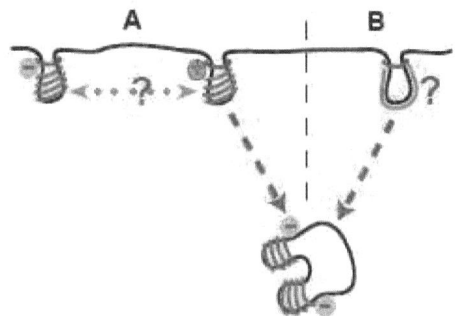

Figure 2-3: Model for caveolae endocytosis
Two models are presented which do not exclude each other. **A**, caveolae consist of two different pools which are either static (-) or endocytically competent (+). How these two pools correspond with one another is not known. **B**, caveolar structures which do not contain caveolae are able to bud from the plasma membrane and fuse with early endosomes or caveosomes. Caveolin expression prevents budding of these structures (from Nichols, 2003).

2.3.4 Role of lipid rafts and caveolae in bacterial infections

Many different bacterial pathogens are reported to subvert lipid rafts and caveolae as entry ports for invasion of the host cell. *E. coli*, one of the first pathogens shown to invade host cells via lipid rafts, attaches to these membrane structures by binding the GPI anchored protein CD48 of macrophages with the pilus tip-located adhesin FimH (Baorto et al., 1997; Shin et al., 2000). This mechanism is thought to depend on the opsonization status of the bacterium, since other pathogens opsonized with antibodies are taken up via clathrin coated pits and are degraded in lysosomes, whereas non-opsonized pathogens circumvent degradation and survive intracellularly. In bladder epithelial cells, FimH promotes adhesion and invasion by recognizing uroplackin-1a, which partitions with the lipid raft membrane fraction. Moreover, Duncan and colleagues show that cholesterol depletion inhibits *E. coli* invasion and caveolin-1 silencing by means of RNAi markedly decreases the ability of the bacterium to invade bladder epithelial cells (Duncan et al., 2004). This observation is somewhat contradictory to the aforementioned data showing that caveolae are stable structures at the plasma membrane in epithelial cells (Thomsen et al., 2002) and that caveolin-1 rather acts as an inhibitor of lipid raft endocytosis by

Introduction

stabilizing lipid rafts at the plasma membrane (Le *et al.*, 2002). Nevertheless, Duncan and colleagues also report that the ability of another Gram-negative pathogen, *Pseudomonas aeruginosa*, to invade alveolar epithelial cells is decreased after RNAi-mediated downregulation of both, caveolin-1 and caveolin-2 expression (Zaas *et al.*, 2005). In addition, the same group observes a decrease in *Pseudomonas* invasion after cell treatment with the tyrosine kinase inhibitor genistein and an increase in invasion due to the application of the phosphatase inhibitor okadaic acid. Genistein is also reported to block uptake of *Campylobacter jejuni* by the enterocyte-like cell line Caco-2, and pretreatment of cells with cholera toxin, which is endocytosed predominantly via lipid rafts (Orlandi and Fishman, 1998; Nichols *et al.*, 2001), decreases intracellular bacteria after *Campylobacter* infection (Wooldridge *et al.*, 1996). Presumably, caveolin-1 is not necessary as a structural component in the endocytic uptake of bacteria by lipid rafts but it is essential as part of the signaling cascade which mediates this uptake.

The presented observations all indicate a positive influence of caveolin-1 in bacterial invasion, and the absence of the protein results in uptake inhibition. In striking contrast to this, recent findings suggest an inhibitory role of caveolin-1 in the invasion of piliated *Ngo* into epithelial cells, which was found to be clustered beneath bacterial microcolonies in infected ME180 cells (Boettcher *et al.*, 2008). This clustering is dependent on a certain caveolin-1 phosphorylation status on Tyrosine 14 since hyperphosphorylated and non-phosphorylated caveolin-1 abolishes its recruitment to gonococcal attachment sites. Non-piliated, Opa-expressing gonococci were unable to recruit caveolin-1 and silencing of caveolin-1 by means of RNAi increased the internalization of bacteria. Complementarily, invasion of AGS epithelial cells, a human gastric carcinoma cell line devoid of caveolin, was inhibited upon transfection with a caveolin-1 construct, and WT invasion levels could by restored by means of RNAi (Boettcher *et al.*, 2008). The blockage of internalization is assumed to allow gonococci the expression of other virulence factors like Opa proteins in order to assure efficient invasion without degradation of the pathogen.

3 Objectives

The aim of this study was to examine host cell factors that are involved in the caveolin-dependent block of piliated, Opa-negative *N. gonorrhoeae* internalization. A previously performed microarray experiment with AGS epithelial cells with and without caveolin infected with *Ngo* revealed a strong upregulation of the PKA-RI subunit in both cell types compared to uninfected cells. To confirm this finding, mRNA from infected cells will be prepared and anylyzed for PKA-RI upregulation by quantative real time polymerase chain reaction (qRT-PCR). In addition, the role of PKA in the infection process of *Ngo*, i.e. the involvement of the kinase in adherence and invasion events of bacteria will be addressed by pharmacological studies and by downregulation of PKA-RI. The third point of interest is the connection between the aforementioned caveolin recruitment by piliated gonococci and the AC/PKA signaling.

4 Results

4.1 Microarray analysis of infected AGS vs infected caveolin-1 expressing AGS cells

In order to assess the influence of caveolin-1 expression in *Ngo*-infected AGS epithelial cells on protein expression, a microarray analysis was performed using AGS cells stably transfected with caveolin-1 and control cells harboring the empty vector. Expression of caveolin was checked by Western blotting and confocal laser scanning microscopy (Fig. 4-1). Table 4-1 and 4-2 show the list of genes which were upregulated more than 4-fold in caveolin-1-expressing and control AGS cells, respectively, after infection with pili-expressing Opa-negative *N. gonorrhoeae* MS11 (strain N280). No gene was found to be down-regulated more than 4-fold in these experiments. In both cell lines, the proinflammatory chemokines IL-8, GRO-alpha (growth-regulated oncogene alpha, Gro1, CXCL1) and GRO-beta (Gro2, CXCL2) were upregulated to a great extent, reflecting the proinfammatory immune response of AGS epithelial cells to gonococcal infection on the mRNA level (table 4-1 and 4-2). While IL-8 is known to be secreted by epithelial cells of the urethra and peripheral blood mononuclear cells upon *Ngo* infection (Ramsey *et al.*, 1995; Lorenzen *et al.*, 1999), the production of CXCL1 and CXCL2 is not described in the literature. Nevertheless, the CXCL1 secretion is observed after challenging AGS cells with *Helicobacter pylori* (Sieveking *et al.*, 2004), and gastric adenocarcinoma cells are thus far not examined for cytokine production upon *Neisseria* infections, explaining the lack of information concerning this issue. Beside its role in immune response, GRO-alpha was identified as a tumor- and angiogenesis-promoting factor in colorectal cancer cells and is constitutively expressed at high levels in melanomas (Luan *et al.*, 1997; Wang *et al.*, 2006). The antiapoptotic factors cIAP-2 (BIRC3) and the RelB-homologue of transcription factor NF-κB (BC028013) involved in tumorigenesis, were also found to be upregulated in the caveolin-negative cells after *Ngo* infection. The cIAP-2 (cellular inhibitor of apoptosis 2) protein prevents apoptosis by binding and inhibiting the cell death proteases caspase-3 and -7 (Roy *et al.*, 1997). In *Chlamydia trachomatis*-infected cells, cIAP-2 is upregulated, leading to the desensitization of these cells to apoptotic stimuli like tumor necrosis factor alpha (TNF-α) (Rajalingam *et al.*, 2006). NF-κB is a well described transcription factor which translocates to the nucleus upon stimulation with TNFα. Like cIAP-2, NF-κB is a potent inhibitor of apoptosis and thereby contributes to the pathogenesis of many human tumors and their chemoresistance. Interestingly, an inhibitor of NF-κB, the zinc finger protein A20 (TNFAIP3) (Beyaert et al., 2000), which concomitantly inhibits apoptosis, shows upregulation in the non-caveolin system, as well as the transmembrane protein lymphotoxin beta, which is an inducer of the inflammatory response system and induced by TNFα (Voon et al., 2004).

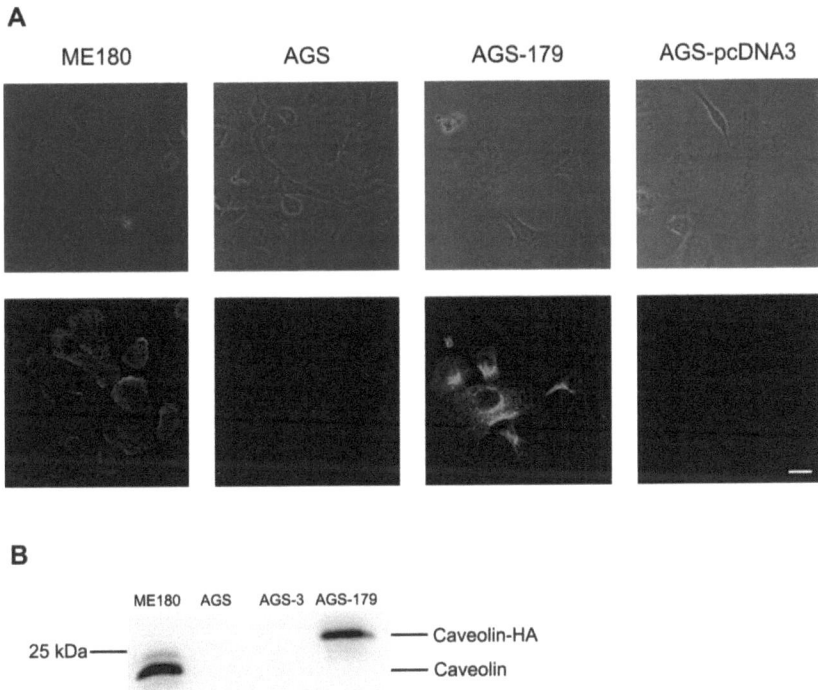

Figure 4-1: Expression of caveolin-1 in epithelial cell lines
A, Expression and distribution of caveolin in ME180, AGS, AGS-179 (stably transfected with pcDNA3-caveolin) and AGS-pcDNA3 cells. Upper panels show cells in phase contrast and lower panel the Cy2-stained caveolin-1 visualized with confocal laser scanning microscopy. Bar, 20 µM. **B**, Immuno detection of caveolin-1 in cell lysates of indicated cell lines. Due to the HA tag, the caveolin-1 protein from stably transfected cells migrates slower than the WT caveolin.

Results

Two more genes were upregulated in both cases: The transcription factor Pax6, which is involved in eye and endocrine pancreas development and the regulatory subunit (RIβ, accession number BC036828) of the cAMP-dependent protein kinase (PKA). In the case of caveolin-expressing cells, RIβ occurred as the strongest upregulated gene (23-fold), while in control cells it appeared as the third most strongly upregulated gene (17-fold). In addition, an upstream factor of PKA, adenylyl cyclase 8 (AC8) also showed up-regulation, but only in infected cells expressing caveolin. The catalytical subunit of PKA is able to interact with caveolin (Razani and Lisanti, 2001) and AC8 colocalizes with caveolae (Schwencke et al., 1999), where it is regulated by capacitive Ca^{2+} entry (CCE) (Smith et al., 2002). Together with the finding that gonococci elicit a calcium influx into infected cells, this data highlights the possibility of a correlation between gonococcal infection, caveolin expression and recruitment and signalling events mediated by the AC/PKA pathway. Thus, it was decided to further examine the role of the PKA signalling pathway in *Ngo*-infected epithelial cells.

4.2 Confirmation of microarray results by quantitative real-time PCR (qRT-PCR)

To verify the upregulation of PKA-RIβ in infected cells, qRT-PCR experiments with mRNA isolated from *Ngo*-infected and non-infected AGS epithelial cells as a control were performed. As shown in Figure 4-2, the PKA-RIβ subunit is upregulated in infected, caveolin-expressing and wild type AGS cells. This is consistent with the findings of the preceding microarray analysis, although the magnitude of up-regulation is more pronounced in this microarray experiment. There, the change in fold magnitude of the RIβ mRNA level was found to be 22.7 [+/-1.1] and 16.8 [+/-1.1] in AGS-caveolin and AGS control cells, respectively, whereas it was found to be 6.8 [+/- 2.0] and 11.3 [+/-5.8] in the qRT-PCR experiment. Despite the difference in magnitude, the qRT-PCR findings confirm the strong upregulation of PKA- RIβ elicited by *Ngo* infection.

Results

Sequence Name(s)	Accession #	Fold Change	Log(Ratio)	P-value	Sequence Description
BC036828	BC036828	22,7009	-1,35604	0,00E+00	Homo sapiens, clone IMAGE:5247772, mRNA, partial cds
IL8	NM_000584	18,28617	-1,26212	9,66E-35	Homo sapiens interleukin 8 (IL8), mRNA.
THC1858902	THC1858902	16,37845	-1,21427	0,00E+00	Unknown
PAX6	NM_001604	14,67326	-1,16653	8,01E-33	paired box gene 6 (aniridia, keratitis)
ENST00000295012	ENST00000295012	10,62803	-1,02645	2,01E-33	Unknown
GRO1	NM_001511	9,34597	-0,97062	0,00E+00	GRO1 oncogene (melanoma growth stimulating activity, alpha)
THC1964042	THC1964042	8,53086	-0,93099	8,80E-18	Unknown
ITGA10	NM_003637	6,44723	-0,80937	3,50E-14	Integrin, alpha 10
HUNK	NM_014586	6,33648	-0,80185	1,94E-12	Hormonally upregulated neu tumor-associated kinase
SORCS3	NM_014978	6,18324	-0,79122	1,11E-26	Homo sapiens sortilin-related VPS10 domain containing receptor 3 (SORCS3), mRNA
BC033035	BC033035	6,16422	-0,78988	2,79E-15	Homo sapiens cDNA clone IMAGE:4839231, partial cds
ADCY8	NM_001115	5,62667	-0,75025	7,51E-08	adenylate cyclase 8 (brain)
NM_152270	NM_152270	4,83807	-0,68467	1,56E-17	Homo sapiens hypothetical protein FLJ34922 (FLJ34922), mRNA
BC028013	BC028013	4,45003	-0,64836	0,00E+00	Homo sapiens v-rel reticuloendotheliosis viral oncogene homolog B, mRNA (cDNA clone MGC:39970 IMAGE:5215944), complete cds
GRO2	NM_002089	3,86717	-0,58739	0,00E+00	GRO2 oncogene

Table 4-1: Gene upregulation in *N.gonorrhoeae* infected AGS-pcDNA3 cells
AGS cells stably transfected with the pcDNA3 expression vector were infected with piliated *N. gonorrhoeae* for 2h and cellular mRNA was isolated and subjected to microarray analysis. Fold change of mRNA content is relative to non-infected control cells. Genes with a more than 4-fold upregulation are displayed.

Results

Sequence Name	Accession #	Fold Change	Log(Ratio)	P-value	Sequence Description
IL8	NM_000584	32,88884	1,51705	0,00E+00	Homo sapiens interleukin 8 (IL8), mRNA.
GRO1	NM_001511	18,16989	1,25935	0,00E+00	GRO1 oncogene (melanoma growth stimulating activity, alpha)
BC036828	BC036828	16,78741	1,22498	0	Homo sapiens, clone IMAGE:5247772, mRNA, partial cds
BC028013	BC028013	8,6211	0,93556	0,00E+00	Homo sapiens v-rel reticuloendotheliosis viral oncogene homolog B, mRNA (cDNA clone MGC:39970 IMAGE:5215944), complete cds
THC1858902	THC1858902	7,26894	0,86147	3,21E-17	Unknown
BIRC3	NM_001165	6,70828	0,82661	1,54E-21	baculoviral IAP repeat-containing 3
PAX6	NM_001604	6,60598	0,81994	4,74E-13	paired box gene 6 (aniridia, keratitis)
GRO2	NM_002089	5,87862	0,76928	1,89E-33	GRO2 oncogene
THC1964042	THC1964042	4,95914	0,69541	4,66E-05	Unknown
TNFAIP3	NM_006290	4,49829	0,65305	2,80E-45	tumor necrosis factor, alpha-induced protein 3
LTB	NM_002341	4,23986	0,62735	0,00E+00	lymphotoxin beta (TNF superfamily, member 3)
AK057151	AK057151	4,11111	0,61396	5,75E-32	Homo sapiens cDNA FLJ32589 fis, clone SPLEN2000443

Table 4-2: Gene upregulation in *N. gonorrhoeae* infected AGS-179 cells
Microarray analysis corresponding to the experiment presented in table 4-1, but with mRNA from infected caveolin-expressing AGS-179 cells stably transfected with pcDNA3-caveolin-HA.

Results

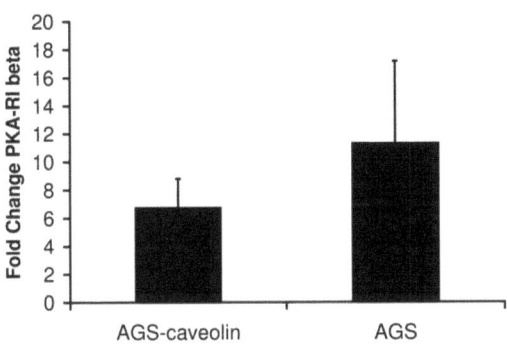

Figure 4-2: Fold change of PKA-RIβ mRNA after *Ngo* infection in caveolin-expressing AGS and AGS control cells
AGS cells stably transfected with caveolin-1 (AGS-caveolin) or AGS control cells were infected with piliated *Ngo* N280. The mRNA content of PKA-RIβ was then determined with quantitative real time PCR (qRT-PCR) and compared with that of uninfected cells.

4.3 Impact of stimulation and inhibition of PKA activity on *N. gonorrhoeae* internalization

PKA is a cytosolic protein kinase which plays a role in many different cellular pathways and functions, which range from such different aspects as glycogen metabolism to gene transcription. It is also involved in the regulation of the cytoskeleton through actin-based processes which are driven by members of the Rho family of small GTPases, integrins and Ena/VASP family members. These proteins are either directly phosphorylated by PKA or regulated by indirect mechanisms as was shown for Cdc42 (Feoktistov *et al.*, 2000; Ellerbroek *et al.*, 2003).

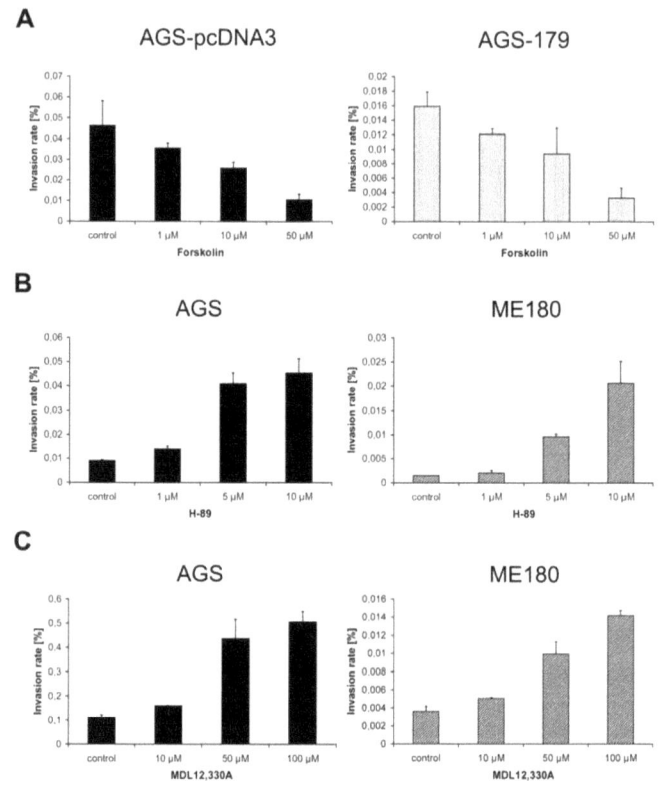

Figure 4-3: Influence of the PKA agonist forskolin and antagonist H-89 on *Ngo* internalization

Different epithelial cell lines were treated with the indicated inhibitor 1h prior to infection with *Ngo* N280 for 2h. **A**, AGS cells expressing caveolin (AGS-179) and AGS control cells (AGS-pcDNA3) were treated with forskolin at given concentrations. The numbers of intracellular and adherent bacteria were determined by a gentamicin protection assay described in Section 8. Values are the ratio of intracellular to adherent bacteria. **B**, Treatment of AGS and ME180 cells with the PKA inhibitor H-89. Samples were processed as in A. **C**, treatment of AGS and ME180 cells with the AC inhibitor MDL12,330A.

Actin rearrangement is a well described feature of infections of epithelial cells with piliated gonococci. As mentioned above, actin cytoskeleton remodelling is essential for pilus-dependent uptake of *Ngo* (Song et al., 2000), and actin accumulation is observed beneath gonococcal clusters attached to epithelial cells (Merz and So, 1997) or around single pilus-negative, Opa-expressing gonococci (Billker et al.,

2002). Gentamicin protection assays were carried out to elucidate the involvement of PKA in the internalization process of gonococci into epithelial cells with or without caveolin. Prior to infection with bacteria, cells were either treated with the PKA activating drug forskolin or the specific PKA inhibitor H-89. As depicted in Figure 4-3A, application of forskolin diminished bacterial internalization into both, caveolin-expressing and control AGS cells. The opposite effect was observed after pre-treatment of AGS and ME180 cells with H-89, which led to a marked increase of gonococcal uptake (Fig. 4-3B). In both cases, the effect was dose-dependent. A prominent difference in the level of internalized bacteria was found in cells without caveolin compared to cells expressing caveolin, as it has been observed earlier (Boettcher et al., 2008). However, as the effects of forskolin and H-89 are monitored in both, caveolin-expressing and non-expressing cells, no correlation was found between the presence or absence of caveolin and the different PKA activation levels regarding gonococcal uptake. This implies that PKA plays a role in an internalization process which is independent of caveolin.

H-89 is hypothesised to be a specific inhibitor of PKA (Leemhuis et al., 2002). However, recent findings reveal that other kinases are also affected by H-89 treatment, although most of them have a higher IC_{50} value for H-89 inhibition (Lochner and Moolman, 2006). One of these proteins which, as well as PKA, is of importance for cellular processes (e.g. signal transduction and membrane dynamics), is PKB/Akt. To rule out the possibility that PKB (and not PKA) or other factors are responsible for the observed effects on Ngo-infected cells treated with H-89, epithelial cells were treated with a compound termed MDL12,330A, which specifically inhibits AC (Delpiano and Acker, 1991), the enzyme which functions directly upstream of PKA in the PKA signalling pathway. Figure 4-3C shows that the effect elicited by the inhibition of AC was similar to that observed with the application of H-89. The number of internalized bacteria rose in a dose-dependent manner, with a 4.6-fold increase with the use of 10 µM MDL12,330A compared to the solvent-treated control for AGS and a 3.9-fold increase in the case of ME180 cells. Taken together, these findings show that the inhibition of the cAMP/PKA pathway, either by inhibiting PKA itself or the upstream factor adenylyl cyclase, results in an increase in the number of internalized bacteria. In addition, the opposite effect is monitored after treatment of the host cell with forskolin, a strong AC activator.

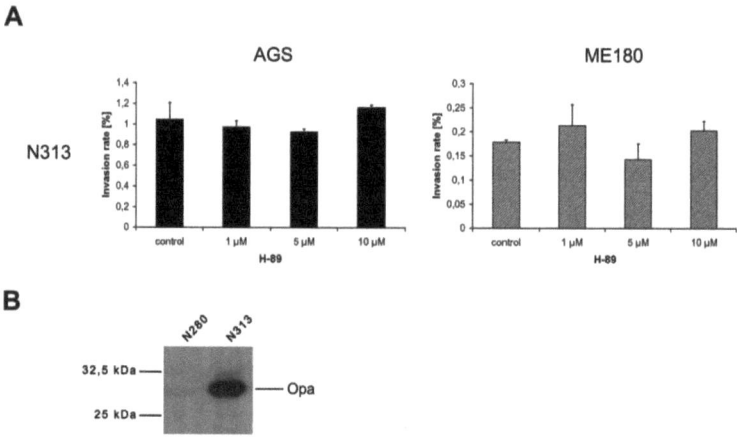

Figure 4-4: Increase in *Ngo* internalization due to PKA inhibition is pilus-dependent
A, AGS and ME180 cells were treated with the PKA antagonist H-89 prior to infection for two hours with none-piliated, Opa-expressing *Ngo* N313. Cells were then subjected to gentamicin protection assay. **B**, Immuno detection of Opa proteins in *Ngo* N280 and N313 lysates with pan-Opa recognizing antibody 4B12/C11.

Pili are an important virulence factor of pathogenic *Neisseria*, because they allow efficient attachment to host cells. Due to their capacity to be retracted, pili guarantee an intimate attachment that allows other factors like Opa proteins, porin and LOS to attach to the cell surface and trigger signalling events that result in the uptake of *Ngo*. To test if the PKA activity level of infected cells influences the invasion of pilus negative, Opa-expressing gonococci, cells were treated with H-89 prior to infection with the gonococcal strain MS11 N313. This recombinant strain expresses the Opa_{57} protein which has been introduced by transformation with a plasmid vector and shows no pili expression due to colony morphology selection (Fig. 4-4B). However, additional adhesion and invasion factors like porin or LOS remain unaltered due to an identical genetic background of the two different MS11 mutants N280 and N313. Invasion level of pilus-negative, Opa-expressing gonococci was not altered by the inhibition of PKA with H-89. In both cell lines used, the amount of invasive bacteria was the same in control cells and in all cells treated with different concentrations of the inhibitor (Fig. 4-4A). This observation strongly argues that pili influence gonococcal internalization not only by mediating bacterial attachment to host cells, but also by signalling events that are based on PKA activation.

4.4 *N. gonorrhoeae* activates the PKA pathway of epithelial cells

As described above, interference with the PKA pathway by pharmacological means changes the pilus-dependent uptake of *Ngo*. Activation of PKA diminished gonococcal internalization, and, vice versa, inhibition of the kinase increased uptake by epithelial cells. These findings show the involvement of PKA in the internalization process, but do not necessarily prove a direct activation of cellular PKA by gonococci. Therefore, epithelial cells were infected with piliated *Ngo* and cAMP levels were determined. Time course experiments with infected ME180 epithelial cells showed increasing concentrations of cAMP within infected cells which peaked at around 30 min post infection and dropped to levels found in uninfected control cells after 2 hours. A slight increase was already monitored after 5-10 minutes after addition of bacteria (Fig. 4-5A). The measured cAMP maximum was about 2.5-fold the value of uninfected cells (Fig. 4-5B). To ascertain that the observed raise in the cAMP level in infected cells is sufficient to activate PKA, the PKA substrate VASP was checked for its phorphorylation status at different time points after infection with piliated *Ngo*. Phosphorylation of VASP results in a slower migration of the protein in SDS-gels and leads to the appearance of a second, 50 kDa signal in addition to the unphosphorylated, 46 kDa protein band. In contrast to mock infected cells, where no alteration of the phosphorylation degree can be observed, cells infected with piliated gonococci elicit a strong increase in the phosphorylation level of VASP (Fig. 4-5C). This finding mirrors the increase monitored for the cAMP concentration shown in Figure 4-5A. Furthermore, like the mock control, cells infected with the pilus-negative, Opa-expressing N313 strain did not trigger VASP phosphorylation (Fig. 4-5C). These results demonstrate that first, PKA is activated in epithelial cells after *Ngo* infection, second, a mild increase in the cellular cAMP concentration triggered by *Ngo* infection is sufficient to activate PKA and, third, PKA activation is pilus-dependent.

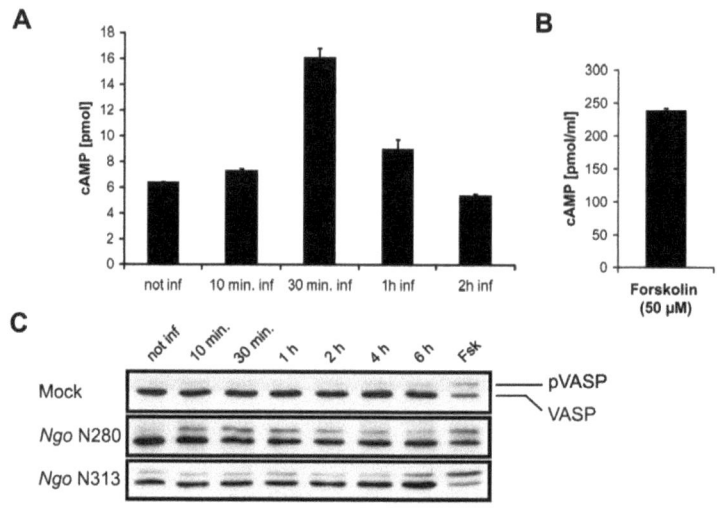

Figure 4-5: Activation of the AC/PKA pathway upon *N. gonorrhoeae* infection
A, The change in cAMP level in ME180 epithelial cells over time after infection with piliated *Ngo* N280. **B**, Increase in cellular cAMP concentration after forskolin (Fsk) treatment. **C**, Phosphorylation of VASP after infection with piliated *Ngo* N280 or Opa-expressing N313. Only infection medium without bacteria was added to the mock samples.

4.5 Recruitment of VASP to infecting *N. gonorrhoeae*

The cellular distribution of VASP within the cell is restricted to sites with high actin dynamics, e.g. the leading edge of lamellipodia and focal adhesions (Reinhard *et al.*, 2001). In the latter case, this localization is due to the interaction of the EVH1 domain of VASP with focal adhesion proteins such as zyxin (Reinhard *et al.*, 1995b) and vinculin (Brindle *et al.*, 1996). This domain is also responsible for the interaction of VASP with the FPPPP motif of the ActA protein of *Listeria monocytogenes*, which targets VASP to the bacterial surface of invasive bacteria. In order to elucidate the localization of VASP in infected ME180 cells, cells were transfected with a VASP-GFP construct, infected with *Ngo*, and analyzed by confocal laser scanning microscopy. Due to the overexpression situation in transfected cells, a considerable fraction of VASP-GFP protein was found in the cytosol. Nevertheless, VASP was also found in structures resembling focal adhesions (Fig. 4-6A), as they were observed in immunfluorescence microscopy samples stained for vinculin (Krause et al., 2003).

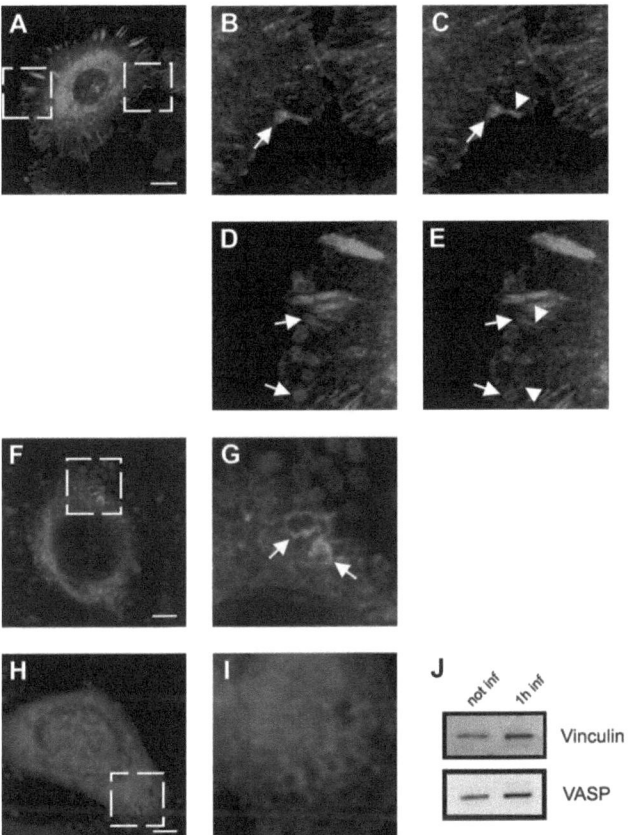

Figure 4-6: VASP recruitment to *N. gonorrhoeae* in VASP-GFP transfected ME180 cells
A-E, Attachment of bacteria (blue) on ME180 cells in close proximity to focal adhesions at the cell periphery. Arrows point to bacteria (blue; B, D) that recruit VASP (green; C, E) and are continuous with straight (C) and wound structures (E) resembling focal adhesions (arrow heads). Bar, 10 µM. **F-G**, Accumulation of VASP (arrows) at the surface of bacteria adhered to perinuclear spaces of a ME180 cell. H-I, Bacteria adhered to a cell expressing GFP. Bars, 5 µM. **J**, Co-immuno precipitation of the focal adhesion marker vinculin with VASP in infected and uninfected ME180 cells. Vinculin was detected in samples precipitated with VASP antibodies (upper panel). Blot membrane was stripped and reprobed with VASP antibodies (lower panel).

In addition, vinculin was co-immunoprecipitated with VASP from ME180 cell lysates (Fig. 4-6J), confirming the co-localization of VASP and focal adhesion structures in the ME180 epithelial cell line. Interestingly, attached gonococci show recruitment of VASP protein which was continuous with focal adhesion structures (Fig. 4-6B, C).

The resulting VASP accumulation was restricted to the size of single diplococci, leaving a VASP-GFP "footprint" on the cellular lamellipodium with a focal adhesion appendage. In some cases, the appendage was not straight like focal adhesions but appeared sidled (Fig. 4-6D, E), raising the question of whether *Ngo* is capable of reshaping focal adhesion structures or has the ability of eliciting *de novo* formation of the observed sidled cellular structure. Bacteria attached in close proximity to the bulk cytosolic VASP-GFP fraction also show VASP recruitment (Fig. 4-6F, G), which does not occur in cells transfected with a GFP construct (Fig. 4-6H, I).

4.6 VASP knockdown affects *N. gonorrhoeae* internalization

As described above, VASP is a key regulator of the actin cytoskeleton. Due to this property and the finding that VASP colocalizes with infecting gonococci (Fig. 4-6), this protein may also be involved in the uptake process of piliated gonococci. This possibility was tested in epithelial cells both with and without caveolin using RNAi. In ME180 cells, bacterial uptake was doubled after efficient knockdown of VASP (Fig. 4-7A). Interestingly, in cells transfected with VASP siRNA the amount of the RIβ-subunit of PKA was also reduced compared to cells transfected with a control siRNA (Fig. 4-7A, left panel). This result is more likely to be caused by an as yet unknown regulation of RIβ by VASP at either the mRNA or protein level than by an off-target effect of the siRNA, because transfection of an additional VASP siRNA distinct in its nucleotide sequence led to the same result. Furthermore, this finding raises the question of whether the observed increase in gonococcal internalization in VASP knockdown cells is more likely an indirect effect of VASP-dependent depletion of RIβ protein than of the knockdown of VASP itself. To rule out which protein was responsible for the observation made in Figure 4-7A (right panel), the same experiment was carried out with a PKA-RIβ specific siRNA. Knockdown of RIβ resulted in an even more pronounced increase in intracellular bacteria compared to the values obtained in the VASP knockdown experiment (Fig. 4-7B). Surprisingly, VASP protein level is also down-regulated by RIβ siRNA transfection, showing a mutual influence of VASP and PKA-RIβ on protein expression of the respective other enzyme. Both knockdown experiments were also performed with AGS cells. In AGS cells that do not express caveolin-1, VASP expression was efficiently abolished and, like in the caveolin-expressing ME180 cells, accompanied by a reduction in the RIβ level (Fig. 4-8A, left panel). However, unlike in ME180 cells, VASP knockdown in AGS cells led to a dramatic decrease in intracellular bacteria (Fig. 4-8A, right panel). Furthermore, AGS transfection with RIβ specific siRNAs did not reduce levels of either protein, RIβ or VASP, compared to control cells (Fig. 4-8B, left panel) and did not influence internalization of gonococci (Fig. 4-8B, right panel). Apparently, RIβ siRNA transfection does not lead to a knockdown of the corresponding protein in AGS cells, although it reliably worked in transfected ME180 cells (Fig. 4-7B, left panel).

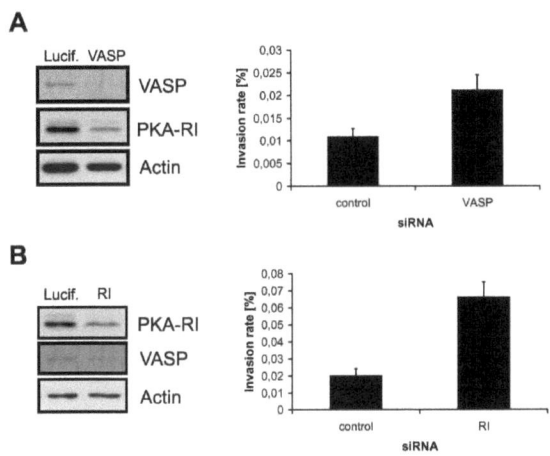

Figure 4-7: Influence of VASP and PKA-RIβ down-regulation in ME180 cells on *N. gonorrhoeae* internalization
A, VASP protein is efficiently down-regulated in ME180 cells treated with a specific VASP siRNA. PKA-RIβ protein level is also significantly decreased (left panel). Internalization is increased in these knockdown cells compared to cells transfected with a control siRNA (right panel). **B**, Down-regulation of PKA-RIβ (left panel). Increase in internalized bacteria in PKA-RIβ knockdown cells is even more prominent compared to VASP-depleted cells (compare B, right panel with A, right panel).

These results show several dramatic differences in the impact of VASP and/or RIβ depletion on gonococcal internalization in cells with and without caveolin. In caveolin-expressing ME180 cells, VASP depletion, in concert with decreased RIβ protein levels, led to an increase in intracellular bacteria, whereas the same changes in protein levels in caveolin-negative AGS cells resulted in dramatically decreased internalization levels. A specific RIβ siRNA is, unlike in ME180 cells, unable to diminish RIβ protein levels in AGS cells. Thus, the possibility that RIβ depletion contributes to the change in intracellular bacteria in AGS cells, as it is monitored in ME180 cells, cannot be ruled out. Another possible explanation for the different outcome of VASP silencing in ME180 and AGS cells with respect to intracellular bacteria may be that internalized bacteria are more susceptible to killing in AGS than in ME180 cells.

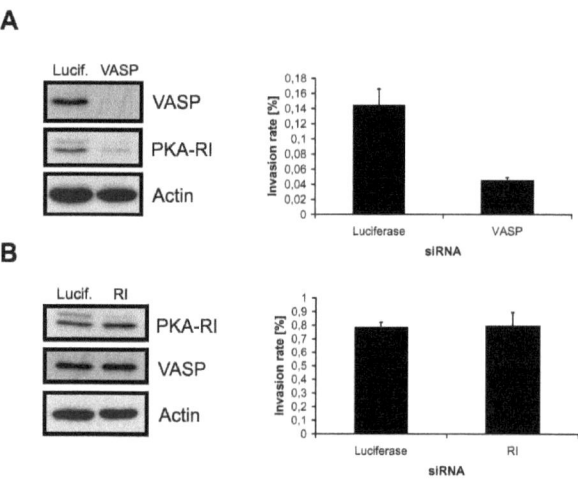

Figure 4-8: Influence of VASP and PKA-RIβ down-regulation in AGS cells on *N. gonorrhoeae* internalization
A, Down-regulation of VASP in caveolin-negative AGS cells leads to a strong decrease in intracellular bacteria. **B**, PKA-RIβ siRNA transfection does not lead to a knockdown of RIβ protein. Protein concentration of VASP stays unaffected (left panel). No change in bacterial internalization was observed (right panel).

Since VASP is an important regulator of actin dynamics, the influence of the actin cytoskeleton-disrupting drug cytochalasin D (cyt D) was investigated under normal and VASP knockdown conditions in ME180 and in AGS cells. Therefore, VASP siRNA transfected cells were treated with 2 µM cyt D before infection with *Ngo* strain N280. For both cell lines, the effects observed in the previous experiments were reproducable. Without cyt D treatment, VASP knockdown led to an increase in bacterial invasion in ME180 and to a decrease in AGS cells. Pretreatment of ME180 cells with cyt D prior to infection did not increase invasion compared to cells transfected with a control siRNA (Figure 4-9A). In contrast, invasion of bacteria into

AGS cells transfected with VASP siRNA is even more reduced on treatment with the actin filament disrupting drug compared to control cells (Figure 4-9B). Apparently, actin filament disruption increases gonococcal invasion into caveolin-1 expressing ME180 cells and concomitantly counteracts the increase of VASP knockdown cells compared to control cells, whereas bacterial invasion into cells devoid of caveolin is inhibited by VASP silencing and even further diminished due to the disruption of actin filaments.

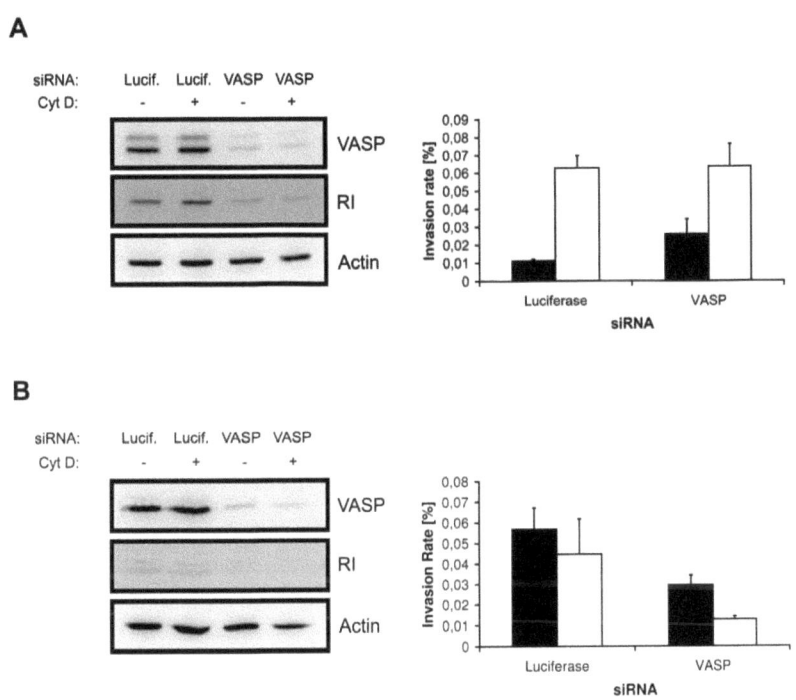

Figure 4-9: Influence of actin cytoskeleton disruption on *N. gonorrhoeae* invasion in VASP knockdown cells
Epithelial cells transfected with VASP siRNA were infected with *Ngo* N280 upon treatment with 2 µM cytochalasin D. **A**, VASP siRNA transfection of ME180 cells results in expected knockdown of VASP and PKA-RIβ (left panel). Cytochalsin D (cyt D) treatment exhibits a strong increase in invasion (white bars) compared to untreated control cells (black bars). Increase in invasion due to VASP/RI downregulation is abolished in cyt D treated cells. **B**, VASP siRNA transfection of AGS cells. Knockdown levels are similar to ME180 transfection. However, cyt D treatment leads to further decrease of bacterial invasion compared to control cells.

4.7 AC/PKA pathway influences *N. gonorrhoeae* association with caveolin and lysosomal compartments

The observations presented in Section 4.3 show that the activity level of the AC/PKA pathway has a strong impact on the internalization of *Ngo*. Pathway inhibition by H-89 increases internalization whereas a dose-dependent block in bacterial uptake is monitored when cells are treated with the AC agonist forskolin.

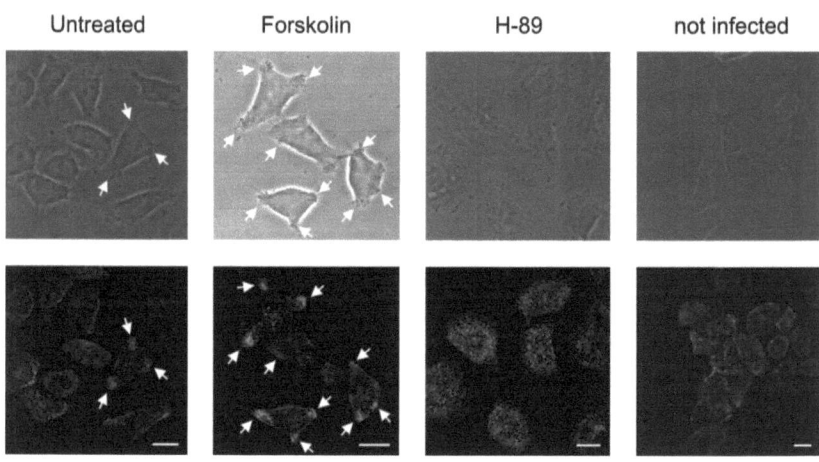

Figure 4-10: Interference with the AC/PKA pathway affects caveolin recruitment
ME180 cells were treated with the AC agonist forskolin, the PKA inhibitor H-89 or were left untreated prior to infection with piliated *N. gonorrhoeae*. Cells were fixed after 1h of infection, stained for caveolin (green) and subjected to confocal laser scanning microscopy. Caveolin-recruiting microcolonies (arrows) are more abundant in forskolin-treated compared to untreated cells and absent in H-89 treated cells. Bars, 15 µM.

Two different hypotheses explain the observed phenomena: first, internalization of bacteria is enhanced by inhibiting the AC/PKA pathway which modifies caveolin and/or actin recruitment or, second, internalization level remains unaltered, but survival of bacteria inside the cell is increased due to this inhibition (e.g. by interfering with the endosomal/lysosomal route of internalization). To test the first hypothesis, ME180 cells were either treated with forskolin or H-89 prior to infection with piliated *Ngo* and stained thereafter for caveolin to verify its recruitment under these conditions by confocal laser scanning microscopy. Forskolin treated cells showed enhanced recruitment of caveolin to bacterial attachment sites compared to untreated control cells, while the opposite effect is observed in H-89 treated cells (Fig. 4-10). Distribution of caveolin in PKA antagonist treated cells appeared similar to that observed in untreated, non-infected cells. This finding shows that influencing the AC/PKA pathway by pharmacological means leads to different caveolin recruitment patterns, which finally result in diminished or enhanced gonococcal internalization. However, the impact of PKA agonists or antagonists on caveolin recruitment to gonococcal microcolonies does not exclude validity of the second hypothesis. Enhanced internalization does not rule out a better survival of bacteria. The combination of both scenarios would have a synergistic effect, rendering it difficult to assess the magnitude of the contribution of the two different effects to the enhanced internalization observed. To test the possibility of improved survival conditions due to AC/PKA pathway inhibition, gentamicin protection assays were performed with MDL12,330A-treated cells for a duration of 2, 3 and 5 hours of gentamicin treatment after *Ngo* infection. As expected, an increase in the levels of internalized bacteria was observed at all time points (Fig. 4-11A). In control cells, the level of intracellular bacteria after 5 hours drops to 10% of the value observed at 2 hours of gentamicin incubation, which indicates the poor survival rate of internalized bacteria. In contrast, compared to the uptake number of non-treated cells, internalization levels showed different dynamics over time with increasing concentrations of the administered AC inhibitor. Bacterial invasion level of cells incubated for 2 hours showed an 4.6-fold increase at a concentration of 100 µM MDL12,330A, whereas a 6.4-fold and even a 13.8-fold increase was monitored at the 3 and 5 hours time points, respectively (Fig. 4-11B). At a concentration of 10 µM MDL, however, an increase of 1.45-fold at 2 hours was observed compared to an increase of only 1.72 at the 5 hour time point. This data shows that blocking the AC/PKA pathway indeed prolongs the survival of internalized gonococci. Hence, increased bacterial internalization is the result of two effects: impaired recruitment of caveolin to gonococcal microcolonies and a better survival of internalized bacteria.

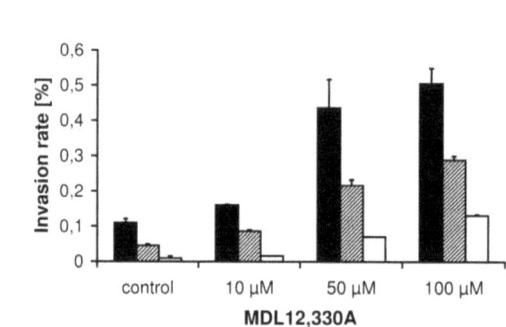

	10 µM	50 µM	100 µM
2h gentamicin	1.45 [+/-0.02]	3.96 [+/-0.73]	4.59 [+/-0.39]
3h gentamicin	1.19 [+/-0.17]	4.8 [+/-0.22]	6.38 [+/-0.25]
5h gentamicin	1.72 [+/-1.2]	7.5 [+/-0.18]	13.79 [+/-0.28]

Figure 4-11: Survival of intracellular *N. gonorrhoeae* is prolonged due to AC/PKA inhibition
Gentamicin protection assays of AGS cells infected with *Ngo* N280 for 2 h with different gentamicin incubation times. MDL12,330A was administered 1h prior to infection at indicated concentrations. **A**, Intracellular bacteria after 2 (black bars), 3 (striped bars) and 5 hours (white bars) of gentamicin treatment. **B**, Relative invasion rate after different periods of gentamicin incubation and MDL12,330A dosages. Numbers are the ratio of invasion rates of cells treated with different MDL12,330A concentrations and control cells at given time points.

A possible explanation for the latter observation is that AC/PKA inhibition interferes with the endocytic pathway. To test this assumption, AGS cells were infected with *Ngo* after treatment with 50 µM MDL12,330A and co-localization of bacteria and the lysosomal marker protein LAMP1 was analysed by confocal laser scanning microscopy. In contrast to untreated cells, colocalization between lysosomal compartments and intracellular bacteria was strongly decreased in cells treated with the AC inhibitor (Fig. 4-12). In some cases, gonococci showed a partial association with LAMP1, which can be interpreted as an escape from the phagocytic vacuole (Fig. 4-12, J-L). Thus, killing of intracellular bacteria was partially prevented by the inhibition of the AC/PKA pathway, probably by disrupting the association of bacteria with lysosomal compartments. In agreement with the gentamicin assay data, there were more intracellular bacteria inside cells treated with MDL12,330A (Fig. 4-12B) compared to untreated cells (Fig 4-12 H).

Results

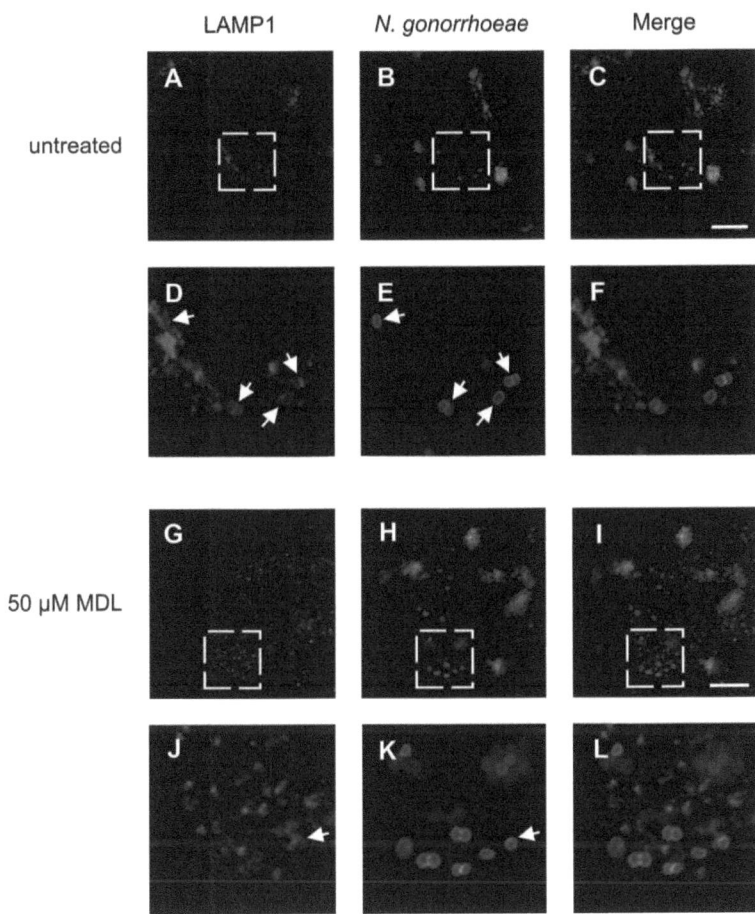

Figure 4-12: AC/PKA pathway inhibition blocks association of intracellular *N. gonorrhoeae* with lysosomes

AGS cells were infected for 2h with *Ngo* 280. Intracellular bacteria appear in green, extracellular in blue/white due to differentially stained bacteria with Cy2 and Cy2/Cy5, respectively (see methods). The lysosomal marker protein LAMP1 appears in red. Bars, 10 µm. **A-F**, AGS control cells with intracellular bacteria associated with lysosomal structures (arrows). **G-L**, AGS cells treated with 50 µM MDL12,330A prior to infection. Arrow marks intracellular bacterium partially associated with lysosomal compartment.

4.8 ErbB2 and Src in *N. gonorrhoeae* infection

Caveolin is phosphorylated at Tyrosin 14 by c-Src (Li *et al.*, 1996), which is a downstream factor of the ErbB signalling pathway. Two members of this receptor tyrosine kinase family, ErbB1 (EGFR) and ErbB2 (Neu) were shown to be recruited to cortical plaques elicited in either *N. gonorrhoeae* or *N. meningitides* infections of epithelial or endothelial cells, respectively (Merz *et al.*, 1999; Hoffmann *et al.*, 2001). In endothelial cells, Src is activated through the *N. meningitides* triggered activation of ErbB2, which in turn leads to invasion of bacteria.

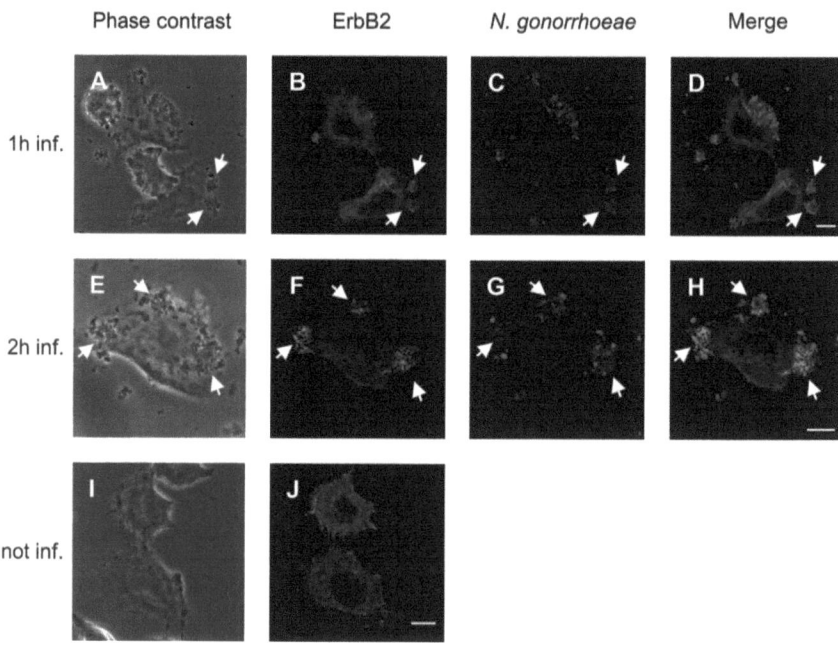

Figure 4-13: ErbB2 recruitment to gonococcal microcolonies
HeLa cells were transiently transfected with N-ErbB2-GFP, a construct which lacks the intracellular portion of ErbB2, and infected with *N. gonorrhoeae* N280. Bars, 10 µm. A-D, Infection of transfected HeLa cells for 1h. ErbB2 (green) is recruited to bacterial microcolonies (arrows). E-H, ErbB2-recruitment after 2h of infection. I-J, None-infected HeLa cells.

To test if ErbB2 is also recruited to cortical plaques caused by *Ngo* infection, HeLa cells were transfected with an ErbB2-GFP construct harbouring the extracellular and transmembrane portion of ErbB2. In uninfected cells ErbB2 is evenly distributed throughout the cytoplasm and concentrated at the plasma membrane (Fig. 4-13I, J).

Results

In contrast, ErbB2 shows a strong recruitment to gonococcal microcolonies in *Ngo* infected cells, which was prominent after one hour and remains stable after two hours of infection (Fig. 4-13A-D, E-H). Furthermore, ErbB2 is linked to the adhesin PilC, because ErbB2 immunoprecipitates with PilC and vice versa (Fig. 4-14). Whether this association is direct or mediated by an adaptor protein (i.e. another component of cortical plaques) could not be distinguished on the basis of this experiment, but it further shows the clustering of bacterial Tfp and cellular components beneath microcolonies.

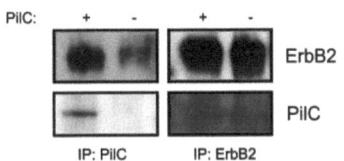

Figure 4-14: Association of pilus subunit PilC with ErbB2
HeLa cells were transfected with ErbB2-WT and incubated with purified PilC protein for 1 h at 4°C. PilC and ErbB2 were immunoprecipitated with PilC serum or a polyclonal anti-ErbB2 antibody, respectively.

Phosphorylation of Src at Tyrosin 416 is an indicator for Src activation level because this modification is necessary for full catalytic activity of the kinase (Brown and Cooper, 1996). Src is not phosphorylated in response to *Ngo* infection compared to uninfected cells, which is also the case for H-89 and forskolin treated cells (Fig. 4-15, upper panel). There was also no detectable increase in caveolin phosphorylation level due to gonococcal infection. However, caveolin-1 phosphorylation was decreased after forskolin treatment (Fig. 4-15, lower panel). Interestingly, two unspecific signals are observed in the Phospho-Src immunoblot, a strong one at approximately 40 kDa and a second, relatively weak signal at around 30 kDA (Fig. 4-15, upper panel). A third unspecific signal is found also in the caveolin immunodetection and migrates at the same level as the lower, unspecific protein in the Src immunoblot (Fig. 4-15, lower panel). These signals suggest that certain cellular factors are tyrosine phosphorylated upon *Ngo* infection that may contribute to internalization processes.

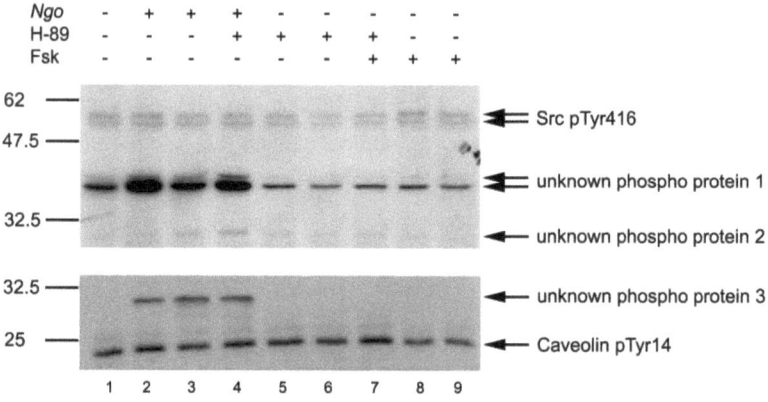

Figure 4-15: Src activity in *N. gonorrhoeae* infected and pharmacologically treated cells
ME180 cells where treated with H-89, forskolin or infected with *N. gonorrhoeae*. In addition, combinations of H-89 and subsequent forskolin treatment/*Ngo* infection were performed. Cell lysates were subjected to Western blotting and probed for phosphor-Src pTyr416 (upper panel) or phosphor-caveolin-1 pTyr14 (lower panel).

4.9 *N. gonorrhoeae* triggered actin and VASP dynamics

Actin plays an important role in attachment and invasion processes of *Neisseria*. Both, *N. meningitides* and *N. gonorrhoeae* were shown to trigger cortical actin rearrangement. These effects depend on either pilus or Opa expression, while cortical plaque formation does not require Opa (Merz and So, 1997; Merz et al., 1999). In agreement with these findings, the piliated, Opa negative strain N280 was able to elicit actin rearrangement (Fig. 4-16) and cortical plaque formation (Fig. 4-13). To reveal the dynamics behind the formation of actin rearrangements, the development of actin recruitment to gonococcal attachment sites was monitored with life cell imaging microscopy. Figure 4-17 shows an actin-GFP transfected cell infected with *Ngo* N280. The onset of actin cluster formation occured as early as 30 minutes after the addition of bacteria. Two different clusters are observed beneath spatially separated bacterial microcolonies, and the merger of these microcolonies during the course of infection is accompanied by the fusion of the associated actin clusters (Fig. 4-17). Actin accumulation reaches a maximum at approximately 1h post infection and declines constantly until 2h post infection when the cluster has almost resolved. Beside the expected actin clusters, actin comets were observed, emanating either from the cytosol prior to and after infection (not shown) or directly from the bacteria triggered actin accumulations (Fig. 4-18).

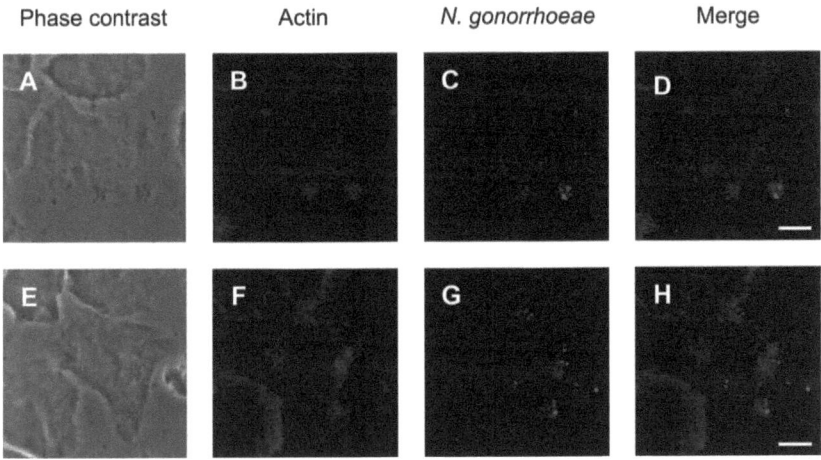

Figure 4-16: Actin recruitment of piliated *N. gonorrhoeae* in infected HeLa cells
HeLa cells were infected with the piliated, Opa-negative (Pil$^+$, Opa$^-$) *Ngo* strain N280, fixed and stained. **A-H**, Two examples of actin (blue) plaque formation beneath bacterial microcolonies (red). Bars, 10 µm.

In addition to actin comets, similar dynamic structures are observed in ME180 and AGS epithelial cells transfected with a VASP-GFP construct. Single comets, which have a life span of up to two minutes, can be followed on their random way through the cytoplasm (Fig. 4-19). Since both, actin and VASP comet structures occur in the absence of bacteria, life cell imaging experiments with fluorescently labelled bacteria are required to determine if these structures are hijacked by internalized bacteria.

Actin comets were already observed 20 years ago by electron microscopy in macrophages infected with the foodborne pathogen *Listeria monocytogenes* (Tilney and Portnoy, 1989). These structures allow the intracellular movement of the pathogen and its spread to neighbouring cells. In more recent studies, similar structures were found to move pinosomes into the cytosol and to target organelles (Merrifield *et al.*, 1999).

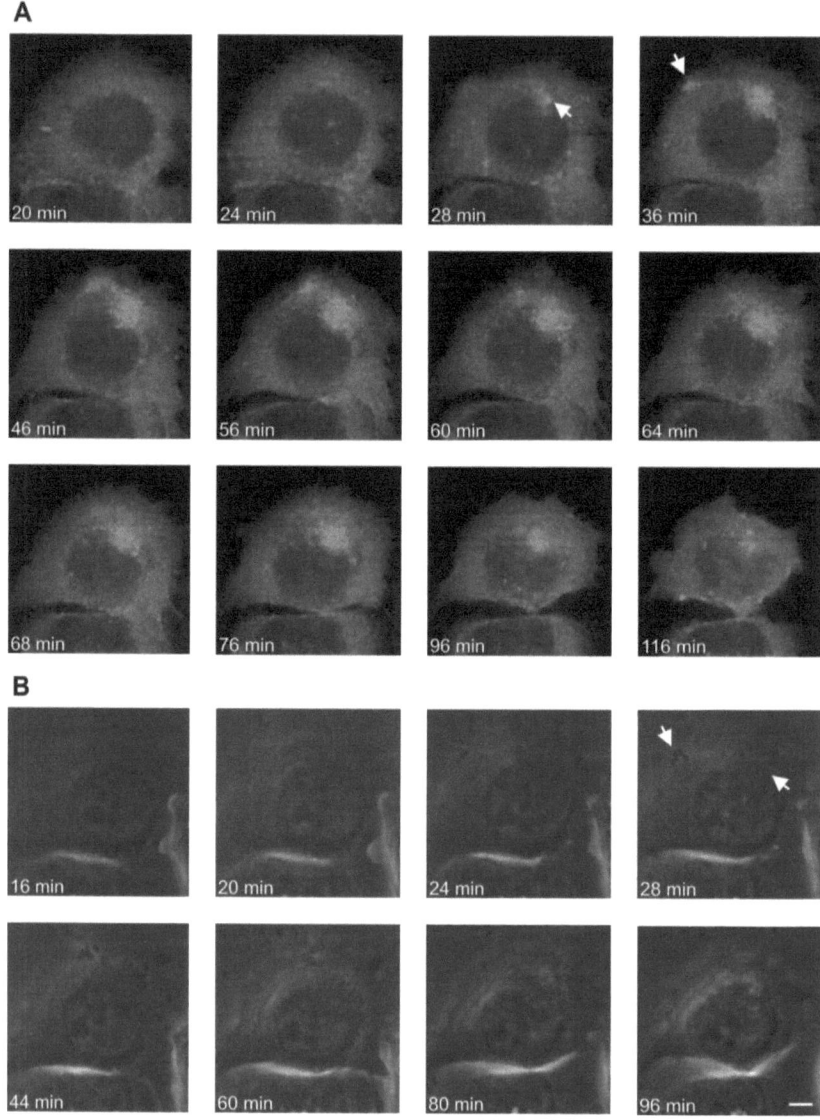

Figure 4-17: Actin cluster formation in epithelial cells infected with *N. gonorrhoeae*
AGS cells were transfected with Actin-GFP and infected with *N. gonorrhoeae*. Actin dynamics were monitored by life cell imaging microscopy. Time points post-infection are indicated. **A**, Actin accumulates beneath two microcolonies (arrows) which fuse with each other. **B**, Corresponding phase contrast images. Bar, 5 µm.

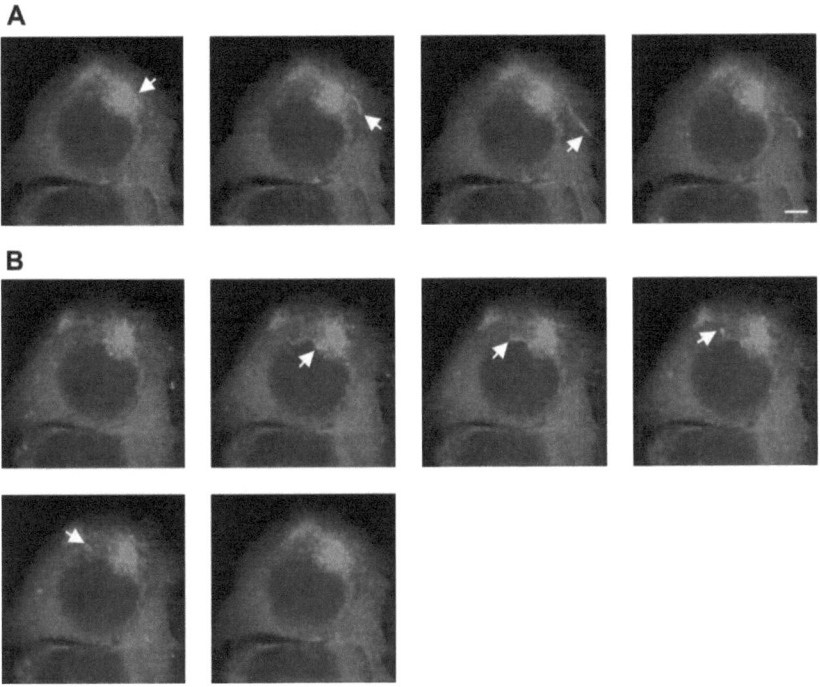

Figure 4-18: Actin comets emanate from *N. gonorrhoaea* triggered actin plaques

A-B, Two examples of actin comets (arrows) leaving the actin accumulation which had developed beneath an attaching bacterial microcolony. Images are taken from successive video images at 25 second intervals. Bar, 5 µm.

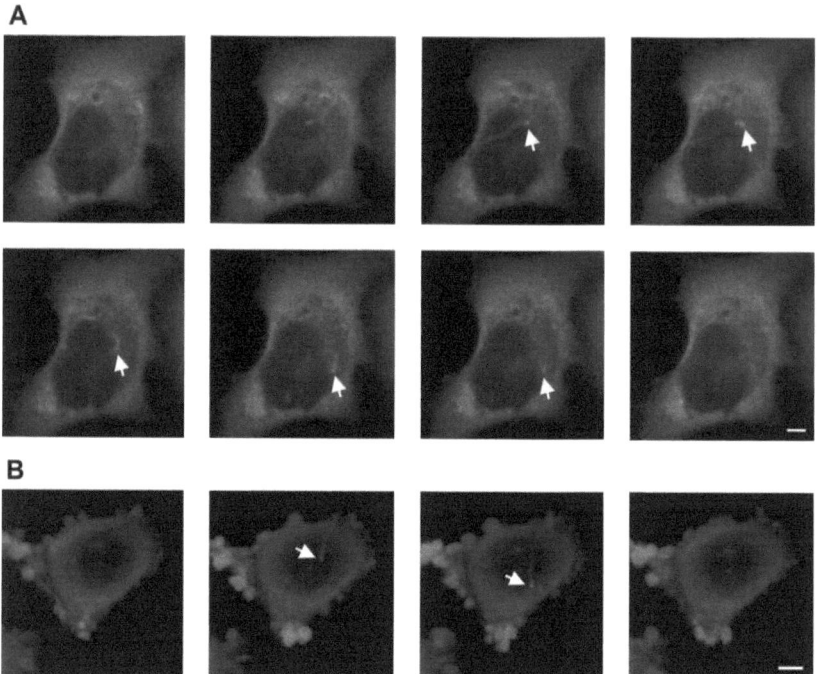

Figure 4-19: VASP comets in epithelial cells
AGS and ME180 cells were transiently transfected with a VASP-GFP construct and subjected to life cell imaging microscopy. Bars, 5 µm. **A**, Trajectory of one comet is followed in an AGS cell over a 200 second period. Images are taken from successive video images at 25 second intervals. **B**, VASP-GFP transfected ME180 cell. Successive pictures are taken at 30 seconds intervals.

5 Discussion

Gonococci have evolved a plethora of factors to adhere to and invade epithelial cells, and their overlapping functions make it difficult to assess the importance of one factor in concert with the others in adhesion and invasion processes. Nevertheless, the function of some gonococcal surface structures like pili, Opa outer membrane proteins and LOS, which are all involved in both, cell adhesion and invasion, have been studied in detail during the last two decades. Although Opa proteins are considered the major invasins, pili provide an important contribution in host cell – microbe interactions that result in bacterial invasion. In meningococcal infections, pili are paramount in adhesion and invasion processes because other factors are masked by the polysaccharide capsule which envelopes the bacterium. Moreover, these findings are established in a primary cell culture model, whereas invasion studies with Opa proteins are usually carried out in transformed cells overexpressing certain Opa receptors. Since pilus negative cells that do not express Opa proteins barely show any adhesion to host cells, it is difficult to examine the role of pili in invasion directly, apart from adhesion events. This might explain why, thus far, pili are not seen as a major factor in gonococcal invasion compared to other bacterial surface components.

5.1 AC/PKA pathway in pilus mediated *N. gonorrhoeae* invasion

In this study, the involvement of the cellular AC/PKA pathway in the *N. gonorrhoeae* infection of cultured epithelial cells was analyzed. The regulatory subunit RIβ of the protein kinase A (PKA) was previously found to be strongly upregulated in caveolin-1 expressing AGS cells upon *Ngo* infection compared to AGS cells devoid of caveolin-1, which was confirmed by means of qRT-PCR.

In several pharmacological studies, the inhibition of either the adenylyl cyclase (AC) with the AC-specific inhibitor MDL12,330A or the PKA of epithelial cells with H-89, resulted in a dosage-dependent increase in gonococcal internalization level. Vice versa, treatment of cells with a PKA agonist, forskolin, induced a strong decrease in intracellular bacteria. Adhesion of gonococci to drug treated cells remained unaffected. Surprisingly, the effects were similar in both cells that express caveolin and those that do not, suggesting that caveolin does not play a role in the invasion

Discussion

events triggered by PKA inhibition. In contrast to this, caveolin recruitment to gonococcal microcolonies is almost completely abolished by PKA inhibition in fluorescence microscopy experiments. Conversely, forskolin strongly promoted the formation of bacteria induced caveolin clusters. These findings show that PKA positively regulates bacteria triggered caveolin recruitment, thereby enhancing the anti-invasive effect of caveolin in gonococcal infection. Since an increase in *Ngo* internalization after PKA inhibition is also monitored in cells that contain no caveolin, PKA might be involved in another uptake process which does not involve caveolin clustering.

PKA is reported to inhibit the small GTPase Rho via phosphorylation at Ser188. This modification inhibits Rho by inhibiting the interaction with its effector ROKα (Dong *et al.*, 1998) or by enhancing the binding of the Rho guanine nucleotide dissociation inhibitor (RhoGDI), which promotes Rho extraction from the cell membrane (Ellerbroek *et al.*, 2003). Moreover, Rac is also inhibited by PKA in endothelial cells, although this is mediated by indirect phosphorylation (Bakre *et al.*, 2002; Ellerbroek *et al.*, 2003). It may therefore be possible that PKA mediated inhibition of small GTPases is responsible for the observed block of internalization in cells devoid of caveolin, because in this work, PKA was found to be activated after *Ngo* infection. In support of this hypothesis, the *Yersinia* protein kinase A (YpkA), a factor which is secreted into the host cell cytosol by pathogenic *Yersinia* strains, is necessary for inhibition of bacterial internalization by affecting the eukaryotic actin cytoskeleton (Wiley *et al.*, 2006; Trasak *et al.*, 2007). Even though there is no information available about bacterial interference with cellular PKA activity yet, these findings show that uptake mechanisms involving small GTPases like Rho can be perturbed via activation of the AC/PKA pathway.

Clearly, the observed enhancement of gonococcal invasion following PKA inhibition and activation of PKA is dependent on the expression of pili. Isogenic Opa-expressing, pilus-negative *Ngo* neither showed an increase in internalized bacteria after H-89 treatment of epithelial cells, nor did they induce PKA activation. Opa proteins are therefore not involved in the internalization process mediated by PKA. That other factors like porin or LOS are also involved in the uptake process described here cannot be ruled out. In previous works, it was demonstrated that gonococci elicit calcium fluxes due to pilus and porin binding (Kallstrom *et al.*, 1998; Muller *et al.*, 1999), with pili triggering calcium release from intracellular stores (Kallstrom *et al.*,

Discussion

1998). These fluxes induce vesicle exocytosis and are believed to enhance *Ngo* intracellular survival (Lin et al., 1997; Ayala et al., 2001; Ayala et al., 2002). Pilus-triggered release of intracellular calcium may have an influence on gonococcal entry into epithelial cells. It is important to note that caveolae play a role in the replenishment of intracellular calcium stores like the ER with calcium, a process which is also referred to as capacitive Ca^{2+} entry (CCE). After release of calcium from the ER, caveolae vesicles move to and fuse with the plasma membrane to mediate calcium refill either by vesicle transport to the ER or by linking plasma membrane and ER directly via Ca^{2+}-channels located in caveolae (Isshiki and Anderson, 2003). Although speculative, this model provides one possible explanation for how *Ngo* invasion is blocked by caveolin recruitment to the site of infection.

In addition to its influence on caveolin recruitment, the AC/PKA pathway has an impact on the survival of intracellular gonococci as shown in Figure 4-11. In relation to the number of internalized bacteria of untreated cells, gonococcal survival is increased if cells are treated with the AC inhibitor MDL12,330A in a dosage-dependent manner. The fold change in the relation of internalized bacteria of treated to untreated cells is 1.45 at 10 µM and 4.59 at 100 µM after 2 hours of gentamicin incubation, whereas the ratio changes from 1.72 to 13.79 at 5 hours. Given that the fold change of the invasion ratios at any time point is the same when cells are left untreated, this data suggests that inhibiting the AC/PKA pathway improves the survival of intracellular bacteria. As shown in fluorescence microscopy experiments with differential staining to distinguish intra- and extracellular bacteria (Fig. 4-12), prolonged survival of intracellular bacteria is due to the dissociation of gonococci and lysosomal compartments. However, compared to the enhanced bacterial internalization after AC/PKA pathway inhibition, the contribution of a better bacterial survival to the increase in intracellular bacteria is only minor.

In previous studies, Opa-expressing gonococci taken up by phagocytic and epithelial cells showed colocalization with LAMP1 (Hauck and Meyer, 1997). In contrast to the highly glycosylated LAMP1 of professional phagocytes, LAMP1 of epithelial cells is reported to be cleaved by the neisserial IgA protease *in vitro* (Hauck and Meyer, 1997) and upon lysosome exocytosis (Ayala et al., 2001). In the experiments presented here, however, the level of LAMP1 in infected cells did not differ to that in uninfected cells. This difference might be due to the different cell lines used, as

Discussion

lysosome exocytosis has been demonstrated for polarized epithelial-like cells (Ayala et al., 2001).

5.2 VASP and *N. gonorrhoeae* infection

ME180 epithelial cells infected with piliated *Ngo* showed a transient increase in VASP phosphorylation at serine 157, a residue phosphorylated by PKA. Moreover, bacteria adhering to VASP-GFP transfected cells exhibited VASP recruitment to their surface, leading to the formation of a VASP-GFP profile in the shape of diplococci. Interestingly, these "footprints" are continuous with VASP tails which can be divided into two subtypes. One subtype is straight and resembles focal adhesion structures, while the other appears sidled and less strong. This evident difference may stem from the fact that the looped pattern results from *de novo* formation due to attaching gonococci, whereas the more prominent, focal adhesion-like structure is preexisting. In the latter case, VASP recruited to bacteria may merge with focal adhesions to form a continuum. VASP is described as a key regulator of actin dynamics (Krause et al., 2003), and this property is likely to be due to its association with the actin monomer-binding protein profilin (Reinhard et al., 1995a) and with monomeric actin itself (Walders-Harbeck et al., 2002). As mentioned above, VASP is phosphorylated at serine 157 by PKA, and this phosphorylation enhances its interaction with actin filaments (Laurent et al., 1999), but has no influence on its subcellular distribution. Thus, elongation of actin filaments triggered by accumulated VASP may result in the formation of the structure observed at gonococcal attachment sites in ME180 cells transfected with VASP-GFP.

In life cell imaging experiments using epithelial cells transfected with either a VASP- or an actin-GFP construct, both proteins showed the formation of comet structures that penetrate the cytosol. In the latter case, comets emanated in part from actin accumulations elicited by adhering *N. gonorrhoeae*, whereas, these structures appeared randomly in the cytosol of uninfected cells. There were no striking differences in size, shape or half life among the comets observed, neither when infected and uninfected, nor when VASP-GFP and actin-GFP-transfected cells were compared. Although described previously in rat fibroblast (Orth et al., 2002), PtK2 (Niebuhr et al., 1993) and Hep2 cells (Gouin et al., 1999), this is the first time that actin and VASP comet tail structures have been observed in ME180 epidermoid and AGS gastric adenocarcinoma cells.

VASP has been shown to be recruited to the surface of *Listeria monocytogenes* by an interaction with the bacterial surface protein ActA (Chakraborty et al., 1995; Gertler et al., 1996). Due to this interaction, it colocalizes with the front of the actin tail comets of intracellularly moving *Listeria*. Although there is no evidence thus far that gonococci are capable of the intracellular locomotion similar to *L. monocytogenes*, this possibility cannot be excluded. There is no information about a gonococcal surface protein that shows homology to ActA, but this is also true for *Rickettsia*. This Gram-negative pathogen also elicits actin-based intracellular movement and recruits VASP to the actin tail front, despite the lack of an ActA homologue. Furthermore, many proteins, like the Arp2/3 complex or ezrin, discovered in the actin tails of *L. monocytogenes* are not found in those of *Rickettsia conorii* (Gouin et al., 1999). This example shows that bacterial actin tail formation is accomplished by different mechanisms, one of which may also be applied by *N. gonorrhoeae*. In EPEC infections, VASP is recruited to the tip of the actin pedestals onto which bacteria stay attached extracellularly for the duration of the infection (Goosney et al., 2000).

Unlike *Listeria* and *Rickettsia*, gonococci are generally thought to be unable to escape the phagosome, which would therefore prevent them from directly interacting with cellular components. However, some observations suggest an occasional escape from the phagosome into the cytosol of epithelial cells (Apicella et al., 1996; Mosleh et al., 1997), and microscopical data from this work also suggest evasion of a lysosomal restriction. Hence, gonococci freely propelling through the cytoplasm is at least an intriguing hypothesis which remains to be confirmed.

5.3 Actin recruitment, dynamics and actin-caveolin interplay

One very important aspect of uptake events during the process of bacterial infections is the reorganization of the actin cytoskeleton. This is either evoked by the pathogen or an intrinsic property of the cell specialized in phagocytosis, as in macrophages or other professional phagocytes. Conversely, actin cytoskeleton modulations are also a feature of bacteria that show little or no invasion of their host cells.

In the case of *N. gonorrhoeae*, the magnitude of the invasive property is dependent on the expression of the different surface factors mediating internalization. As

Discussion

demonstrated in this work and in previous studies, Opa protein-expressing gonococci devoid of pili exhibit a one order of magnitude higher invasive phenotype than piliated, Opa-negative bacteria. Despite this, both variants are able to accumulate actin beneath their attachment sites and this is not abolished by the actin filament disrupting drug cytochalsin D (Grassme et al., 1996; Merz and So, 1997). In contrast, the impact of cyt D on internalization is completely different. It is reported to efficiently inhibit the uptake of Opa-expressing, pilus negative gonococci by different epithelial cell lines (Grassme et al., 1996; McCaw et al., 2003), whereas experiments with piliated Ngo revealed an up to 5-fold increase in the uptake rate by cyt D-treated ME180 epithelial cells (Boettcher et al., 2008). Application of the actin monomer-sequestering drug latrunculin A (Lat A) caused an even more pronounced gonococcal uptake. Surprisingly, cyt D-treatment of AGS cells resulted in substantial decrease of gonococcal internalisation (Fig. 4-9). These findings may reflect that the presence of both, caveolin-1 and an intact actin cytoskeleton, is crucial for pilus-mediated internalization of Ngo. Caveolae are anchored to the actin cytoskeleton via the F-actin crosslinking protein filamin, which additionally binds to caveolin-1 and therefore contributes to the spatial stability of caveolae. In the case of actin cytoskeleton disruption by cyt D, anchorage is abolished, caveolae become motile and allow the observed gonococcal entry into ME180 cells. Indeed, inhibition of filamin expression by means of RNAi resolves the caveolin-1 clusters beneath microcolonies and induces gonococcal internalization (Boettcher et al., 2008). In support of the these findings, actin filament depolymerization with Lat A induces lateral movement of otherwise stationary caveolae in the plasma membrane (Pelkmans et al., 2002).

Discussion

5.4 Signalling through caveolae/lipid rafts relevant to *N. gonorrhoeae* internalization

In addition to the suggested function of caveolin-1 as a physical barrier that blocks uptake of *Ngo*, caveolar compartments may also provide the basis for signalling events that have an impact on invasion. As indicated above, many cell signalling components are located in the caveolar/lipid raft fraction or at least associate with

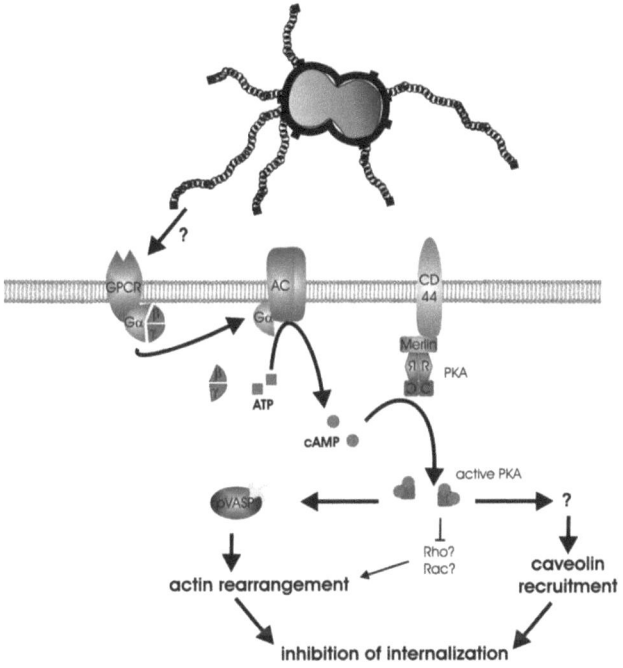

Figure 5-1: Inhibition of *N. gonorrhoeae* internalization into epithelial cells
Activation of PKA leads to caveolin recruitment and actin re-arrangement that cooperatively results in a block of bacterial internalization. The pilus-dependent process is supposedly initiated by activation of a G-protein-coupled receptor (GPCR) rather than by a direct activation of the adenylyl cyclase (AC). Active PKA phosphorylates VASP which in turn triggers actin polymerization. In concert with caveolin-recruitment and possibly by inhibition of the small GTPases Rac and Rho, internalization of gonococci into epithelial cells is blocked. Efficient signalling is assured by compartmentalization of involved factors in lipid rafts/caveolae.

these membrane microdomains. Proteins which accumulate beneath *Ngo* microcolonies are reported to localize to caveolae/lipid rafts, namely CD44/I-CAM1 (Neame *et al.*, 1995), EGFR (Couet *et al.*, 1997b; Matveev and Smart, 2002) and ErbB2 (Nagy *et al.*, 2002). The latter is thought to be activated by pilus-mediated *N. meningitides* adherence through its clustering beneath microcolonies (Hoffmann *et al.*, 2001), which in turn leads to Src as well as cortactin activation and finally results in actin cytoskeleton remodelling and bacterial entry. Cortactin recruitment to the plasma membrane is achieved by LOS-dependent activation of the PI3-kinase and Rac (Lambotin *et al.*, 2005). Analogous to *N. meningitides* infection, ErbB2 is also clustered beneath gonococcal microcolonies and seems to be associated with the pili since ErbB2 was immunoprecipitated with the pilus subunit PilC and vice versa.

In addition, pili and LOS interactions with specific cell surface receptors are both involved in Opa-independent internalization of gonococci (Song *et al.*, 2000), and elongation of epithelial microvilli induced by *Ngo* requires LOS interaction with the asialoglycoptotein receptor (Harvey *et al.*, 2001). However, there is no data on cortactin recruitment to gonococcal microcolonies and, in contrast to *Nmg* infection, Src activation was not observed during *Ngo* infection. Consequently, in the presence of caveolin-1, invasion of gonococci is efficiently blocked, resulting in a two orders of magnitude lower bacterial invasion compared to meningococcal invasion into endothelial cells, which express high levels of caveolin-1. *Nmg* shows a intracellular/adherent bacteria ratio of 1.3% (Lambotin *et al.*, 2005), whereas a ratio of 0.005% was observed for *Ngo* infection of ME180 cells. Although no direct activation of EGFR and/or ErbB2 signalling was detected via the phosphorylation of Src in infected cells, stimulation of cells with EGF prior to infection leads to enhanced Src-dependent phosphorylation of caveolin-1 and increased invasion (Boettcher *et al.*, 2008). Complementarily, inhibition of Src reduces EGF-induced bacterial internalisation, and a phosphorylation defective mutant of caveolin-1 (Cav1-Y14F) is not recruited to *Ngo* microcolonies and does not prevent bacterial internalisation into Cav1-Y14F transfected cells (Boettcher *et al.*, 2008). Thus, a certain degree of caveolin phosphorylation is important for its recruitment and inhibition of gonococcal entry. Since other tyrosine phosphorylation events are detected upon *Ngo* infection, gonocooci may directly influence their internalization via other signalling components. Besides the aforementioned cortical plaque components, basically all required factors involved in AC/PKA signalling are localized to lipid rafts/caveolae. In detail, raft

Discussion

localization was reported for β-adrenergic receptors (Schwencke et al., 1999), $G_{\alpha s}$-subunit of G-protein coupled receptors (Ostrom et al., 2002) and adenylyl cyclase type 8 (Smith et al., 2002). For PKA, interactions with caveolin-1 (Razani and Lisanti, 2001) and ERM-family proteins like ezrin (Dransfield et al., 1997) and merlin (Gronholm et al., 2003) have been demonstrated. Merlin, which binds to the raft-located membrane protein CD44/ICAM-1 (Sainio et al., 1997), was shown to specifically interact with PKA-RIβ (Gronholm et al., 2003). These findings suggest that caveolin recruitment elicited by piliated gonococci may arrange AC/PKA signalling components in close proximity to each other (Fig. 5-1). Subsequently, bacteria block their internalization by actin modulations through PKA activation that stabilizes the caveolin-actin barrier. Recruited caveolae are efficiently crosslinked by filamin. The latter is a substrate of PKA, and phosphorylation of filamin enhances its crosslinking capability (Jay et al., 2000). Additionally, PKA-mediated VASP phosphorylation, which strongly enhances its affinity for filamentous actin (Laurent et al., 1999), may promote actin polymerization, thereby contributing to the caveolin-actin barrier setup at the attachment site of gonococci (Fig. 5-2). Inhibition of PKA, by contrast results in impairment of caveolin recruitment and actin polymerization, weakening of the caveolin-actin meshwork and subsequent enhancement of bacterial internalisation.

Caveolin clustering is also promoted by activated PKA. PKA activation resulted in potentiated caveolin clustering beneath microcolonies, whereas its inhibition prevented caveolin clustering (Fig. 4-10). The mechanism how PKA influences caveolae/lipid raft mobility is not known, but clustering of caveolin might be achieved by pilus binding to a factor localized in caveolae/lipid rafts. Subsequent retraction of pili then accumulates caveolae and activated PKA associated to these microdomains assures their anchorage beneath bacteria. A component of the AC/PKA pathway, the G-protein coupled receptor (GPCR), is a transmembrane protein which owns an extracellular portion accessible to pili. It is therefore possible that PKA activation may occur via pili-dependent attachment and subsequent activation of a certain member of the GPCR family. Since Ngo does not secrete toxins like Vibrio cholerae, whose toxin constitutively activates the AC, it is unlikely that PKA activation is accomplished via pilus-mediated activation of the AC. Moreover, AC activation is assured by association of the $G_{\alpha s}$ subunit with the intracellular fraction being protected from gonococcal factors by the plasma membrane.

Discussion

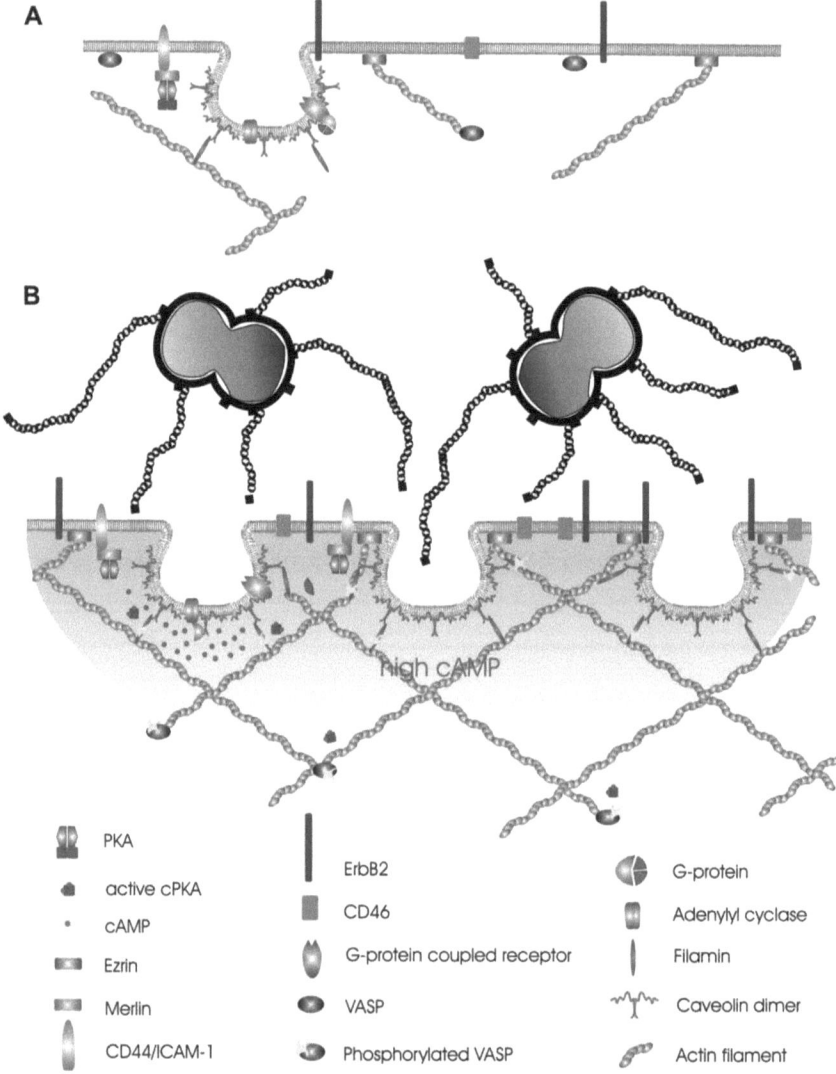

Figure 5-2: Model for *N. gonorrhoeae* induced caveolin- and actin clustering
A, PKA coupled to the plasma membrane is inactive in none-infected cells. **B**, Upon infection, gonococci induce cAMP production and subsequently activated PKA phosphorylates VASP. VASP then polymerizes actin beneath *Ngo* attachment sites and is dephosphorylated by phosphorylases in more distant cytosolic compartments where cAMP concentration is low, restricting VASP-promoted actin polymerization to areas in close proximity to attaching bacteria. Connection between recruited caveolin and actin by filamin is enhanced by filamin phosphorylation, which is also facilitated by PKA.

6 Conclusion

The cAMP-dependent protein kinase (PKA) and the upstream factor adenylyl cyclase (AC) are involved in piliated, Opa-negative *N. gonorrhoeae* invasion. They are activated upon infection and contribute to actin polymerization via VASP activation. Actin polymerization and caveolin recruitment to bacterial attachment sites, which is also influenced by PKA, both contribute to the block of invasion. Silencing of PKA-RI or VASP both resulted in increased gonococcal internalization in ME180 epithelial cells, whereas caveolin-deficient AGS cells revealed decreased uptake. In addition to the AC/PKA pathway, other as yet unknown factors also account for the block of invasion. VASP exhibited colocalization with gonococcal attachment sites on infected cells and was linked to VASP structures resembling comet tails that intrinsically occur in epithelial cells. Upregulation of the PKA subunit RI was confirmed by qRT-PCR. Whether this effect is triggered by infecting bacteria or by the cell counteracting infection remains to be addressed.

7 Materials

7.1 Bacteria

7.1.1 E. coli

Strain	Genotype
DH5α	F'/endAl hsdR17 (rk-mk+) sup E44 thi-1 recA1 gyrA (Nalr) relA1 (lacZYA-argF) U169 (80dlac(lacZ) M15
TOP10	F'{lacqTn10 (TetR)} mcrA D (mrr-hsdRMS-mcrBC) φ80lacZΔM15 ΔlacX74 deoR recA1 araD139 Δ(ara-leu) 7697 galU galK rpsL endA1 nupG

7.1.2 N. gonorrhoeae

Strain	properties
MS11 N280	MS11-F3-54b; *opaC::cat*; P$^+$, Opa $^-$
MS11 N313	MS11-B1-87; *opaC::cat*; P$^-$, Opa $^+$; *Opa$_{57}$* in pTH6a plasmid
MS11 N560	P$^-$, Opa $^-$; Δ*pilE1 pilC1::ermC' pilC2::cat*$_{low}$; *pilC2*$_{His6}$^{iv} in Hermes-8 plasmid

7.2 Cell culture

Cell line	Specification	Additional properties	ATCC no.
AGS	human gastric adeno-carcinoma		CRL-1739
AGS179		stably transfected with pcDNA3-Cav1-HA	
AGS pcDNA3		stably transfected with pcDNA3	
ME180	cervix epidermoid carcinoma		CRL-7932
HeLa	cervix adenocarcinoma		CCL-2

7.3 Cell culture media and supplements

Cell line	Medium	Supplier
AGS	RPMI 1640	Gibco
AGS cav 179	RPMI 1640	Gibco
AGS pcDNA3	RPMI 1640	Gibco
ME180	McCoy's 5A Medium	Gibco
HeLa	RPMI 1640	Gibco

Solution	Supplier
D-PBS	Gibco
FCS	Biochrom AG
Geneticin (G418)	Invitrogen
Lipofectamine^{TM} 2000	Invitrogen
Optimem	Invitrogen
Trypsin-EDTA solution	Invitrogen

7.4 Media for bacterial culture

Medium	Composition
LB broth	10 g bacto-trytone, 5 g yeast extract, 5 g NaCl in 1 L ddH$_2$O
LB agar plates	LB broth with 1.5% agar
GC agar plates	7.5 g caseine peptone, 7.5 g meat peptone, 1 g KH$_2$PO$_4$, 4 g K$_2$HPO$_4$, 1 g amylomaize, 5 g NaCl, 10 g Agar in 1 L ddH$_2$O; after autoclaving, 1% vitamin mixture is added
SOC medium	20 g bacto-tryptone, 5 g yeast extract, 0.5 g NaCl, 2.5 ml KCl (1 M) in 1 L ddH$_2$O; adjust to pH 7.0; after sterilization by autoclaving, 20 ml sterile glucose (1 M) and 5 ml MgCl$_2$ (2M) are added
PP medium	15 g proteose peptone, 5 g NaCl, 0.5 g

Materials

Medium	Composition
	amylum, 4 g KH_2PO_4, 1 g K_2HPO_4 in 1 L ddH_2O; adjust to pH 7.5
Skim milk	10 % low-fat milk powder in ddH_2O
Vitamine mixture	0.01 g vitamine B12, 1 g adenine, 0.03 g guanine, 10 g glutamine, 0.03 g thiamine, 25.9 g L-cysteine, 1.1 g L-cystine, 0.5 g uracil, 0.15 g arginine, o.o2 g $Fe(NO_3)_3$, 0.013 g p-amino benzoic acid, 0.1 g cocarboxylase, 0,25 g nicotinamide adenine dinucleotide (NAD), 100 g D-glucose in 1 l ddH_2O

7.5 Plasmid vectors

Plasmid	comments	supplier
pEGFP-N1		clontech
pll3.7	Addgene no. 11795	
pVASP-GFP		kindly provided by Frank B. Gertler
pAcGFP1-Actin		clontech
pErbB2-GFP	ErbB2 extracellular fragment plus transmembrane domain (AA 1-685) in pEGFP-N1	

7.6 Oligonucleotides

siRNAs

Gene	Accession no.	Sequence	Position (bp)
PKA-RIβ	BC036828	CAGGGAGTTGAGGCCGAAGAA	1720-1740
VASP	NM_003370.1	AACTTCGGCAGCAAGGAGGAT	528-548
VASP-2	NM_003370.1	CCGGGCCACTGTGATGCTTTA	281-301
VASP-3	NM_003370.1	TAGATTCACTTTAACGCTTAA	1726-1746
Luciferase	M15077	AACTTACGCTGAGTACTTCGA	513-533

Primers for qRT-PCR

Gene	Accession no.	Sequence	Position (bp)
PKA-RIβ	BC036828	ACAGCAAGGGAATGAAAGAG	601
		TCCATCTTCAAACTGGACGG	924
GAP-DH	NM_002046	GGTATCGTGGAAGGACTCATGAC	531
		ATGCCAGTGAGCTTCCCGTTCAG	718

7.7 Antibodies

Primary antibodies

Antibody	Origin	Dilution	Supplier	Order no.
β-Actin	Mouse	1:5000 (IB)	Sigma-Aldrich	A5441
Caveolin		1:1000 (IB)	Santa Cruz	sc-894
		1:100 (IF)		
Caveolin pTyr14	Rabbit	1:1000 (IB)	Santa Cruz	sc-14037
ErbB2	Maus	1,25 µg/ml	Oncogene	OP15-100UG
LAMP1	Maus	1:100 (IF)	Southern Biotech	9835-01
PKA-RIβ	Maus	1:100 (IB)	BD Transduction	610165
c-Src pTyr416	Rabbit	1:1000 (IB)	Cell Signaling	2101
c-Src pTyr527	Rabbit	1:1000 (IB)	Cell Signaling	2105
VASP pSer157	Rabbit	1:1000 (IB)	Cell Signaling	3111
VASP	Mouse	1:1000 (IB)	BD Transduction	610447
VASP	Mouse	IP: 2 µg/sample	Santa Cruz	sc-46668
Vinculin	Maus	1:1000 (IB)	Millipore	CBL233
N. gonorrhoeae (AK213)	Rabbit	1:100 (IF)	MPI for biology	
N. gonorrhoaea	Rabbit	1:50 (IF)	U.S. Biological	NO600-02
PilC (rabbit serum)	Rabbit	1:100 (IF)	MPI for biology	

Secondary antibodies

Antibodies Immunoblot (IB)	Dilution	Supplier
ECL™ donkey anti-rabbit IgG, HRP-linked	1:3000	GE Healthcare
ECL™ sheep anti-mouse IgG, HRP-linked	1:3000	GE Healthcare

Antibodies Immunofluorescence (IF)	Dilution	Supplier
Cy^{TM}2-conjugated goat anti-rabbit IgG	1:150	Jackson ImmunoResearch
Cy^{TM}2-conjugated donkey anti-mouse	1:150	Jackson ImmunoResearch
Cy^{TM}3-conjugated goat anti-rabbit IgG	1:150	Jackson ImmunoResearch
Cy^{TM}3-conjugated donkey anti-mouse	1:150	Jackson ImmunoResearch
Cy^{TM}5-conjugated goat anti-rabbit IgG	1:150	Jackson ImmunoResearch
Cy^{TM}5-conjugated donkey anti-mouse	1:150	Jackson ImmunoResearch

7.8 Buffers and solutions

Indirect immunofluorescence

Solution	Composition
4% PFA, pH 7.4	8 g of PFA are dissolved in 100 ml ddH$_2$O, suspension is heated to 65°C and 1 M NaOH is added dropwise under stirring until the solution appears clear. 20 ml 10x PBS are added, pH is adjusted to 7.4 and the final volume of 200 ml is reached by adding ddH$_2$O. Aliquots are stored at -20°C.
Permeabilization	0.1% TritonX-100 in PBS
PBS (Phosphate buffered saline)	15 mM NaCl, 0.84 mM Na$_2$HPO$_4$, 0.18 mM NaH$_2$PO$_4$, pH 7.4
Blocking	1% BSA, 0.05% Tween20 in PBS
Mounting	2.4 g Mowiol 4-88 are dissolved in 7 g glycerol and 6 ml ddH$_2$O under stirring at RT for 2h. The solution is incubated for 1h at 50°C after addition of 0.2 M Tris/HCL (pH 8.5) and then stirred at 50°C until Mowiol is dissolved. To withdraw undissolved Mowiol clumps, the solution is centrifuged at 7500x *g* for 15 min and stored at 4°C.

Gentamicin protection assay

Solution	Composition
Saponin solution	1 % (w/v) Saponin in RPMI 1640, sterile filtrated
Gentamicin solution	50 µg/ml gentamicin in RPMI 1640

SDS-PAGE and Western Blot

Buffer	Composition
5x Sample buffer	5% SDS, 0.5% bromphenol blue, 0.25 M Tris/HCl pH 6.8, 10 mM EDTA, 50% glycerol, 1% β-mercaptoethanol
Running buffer	25 mM Tris/HCl pH 8.3, 192 mM glycine, 0.1% (w/v) SDS
Transfer buffer (Wet blot)	25 mM Tris, 190 mM glycine, 20% methanol
Transfer buffer (Semi dry)	48 mM Tris, 39 mM glycine, 0,035 mM SDS, 20% (v/v) methanol
Blocking solution	10% nonfat dry milk, 3% (w/v) BSA in TBS-T
Wash Buffer (TBS-T)	10 mM Tris/HCl pH 7.5, 100 mM NaCl, 0,1% Tween-20
Stripping Buffer	62.5 mM Tris/HCl pH 6.7, 100 mM β-mercaptoethanol, 2% SDS

Immuno precipitation

Buffer	Composition
Cell lysis buffer	20 mM Tris (pH 7.5), 150 mM NaCl, 1 mM EDTA, 1 mM EGTA, 1% Triton X-100, 2.5 mM sodium pyrophosphate, 1 mM β-Glycerolphosphate, 1 mM Na_3VO_4, 1 µg/ml Leupeptin

Protein staining solution

Solution	Composition
Ponceau S	2% (w/v) Ponceau S, 30% (w/v) TCA, 30% (w/v) sulfosalicylic acid
Coomassie staining solution	025% (w/v) Coomassie Brilliant Blue R250, 50% Methanol, 10% glacial acetic acid
Destaining solution	10% ethanol, 10% glacial acetic acid

PilC purification

Buffer	Composition
PP-medium	1% vitamine mixture, 10 mM MgCl2, 5 mM NaHCO3
P1 (resuspension buffer)	50 mM Tris/HCl pH 8.0, 150 mM NaCl
P2 (extraction/equilibration buffer)	50 mM Tris/HCl 8.0, 500 mM NaCl, 2% LDAO
P3 (wash buffer)	20 mM Imidazole, 10% glycerol, PBS pH 7.4
P4 (wash buffer)	10% glycerol, PBS 7.4
P5 (elution buffer)	1mM citric acid, 10% glycerol, 150 mM NaCl; adjusted to pH 4.0
P6 (total elution buffer)	10 mM EDTA, PBS 7.4

7.9 Chemical reagents

Chemicals which are not listed here were purchased from Roth, Merck, Serva, and Sigma-Aldrich.

Chemical	Supplier
Complete	Roche
ECL substrate solution	PerkinElmer
Forskolin	EMD Biosciences
FCS	Biochrom
GC Agar base	Remel
Glycine	Biomol
H-89	EMD Biosciences
HiPerFect transfection reagent	Qiagen
ImmunoPure Immobilized Protein A/G	Pierce
LB Agar base	Invitrogen
Meat peptone	SIFIN
MDL12,330A	EMD Biosciences
Protein A agarose	EMD Biosciences
Protein G agarose	EMD Biosciences

Materials

Chemical	Supplier
Proteose Peptone	Becton Dickinson
Sucofin low-fat milk powder	TSI
Tris (hydroxymethyl)-aminomethan	AppliChem

7.10 Kits

Name	Supplier
Parameter™ Cyclic AMP Assay	R&D Systems
RNEasy kit	Qiagen
Qiafilter™ Midi Plasmid Purification	Qiagen
Nucleobond® AX	Macherey-Nagel

7.11 Appliances and consumable materials

Shaking incubator G25 (New Brunswick Scientific), luminescent image analyzer LAS-3000 (Fujifilm Life Science), Centrifuge 5417C (Eppendorf), Centrifuge 5417R (Eppendorf), Sorvall® RC-5B (Kendro Laboratory products), confocal laser scanning microscope TCS SP (Leica), stereo microscope SZ-60 (Olympus), DIC light microscope IX-50 (Olympus, Hamburg), Mini-PROTEAN® III Electrophoresis Cell (BioRad), Western Blot device Fastblot B33 (Biometra), HERA cell 150 incubator (Heraeus), photometer DR/200 (Hach), spectral photometer UltraSpec 3000 (Amersham Biosciences), Nano drop Spectrophotometer ND-1000 (peqLab Biotechnologie), DUOMAX 1030 shaker (Heidolph), Abi Prism 7900HT (AME Bioscience), ELISA reader Spectra Max 190 (Molecular Devices), OPTIMAX 2010 X-Ray Film Processor (Protec Medizintechnik)

Glass vessels (Schott), Amersham Hyperfilm™ (GE Healthcare), PVDF Transfer Membrane (PerkinElmer), Whatman chromatography paper (Schleicher and Schüll), reaction tubes (Sarstedt), Falcon tubes (Sarstedt), microscope slides (Marienfeld), 12 mm cover slips (Roth), cell culture dishes (TPP and Corning Life Sciences), sterile pipets (Corning Life Sciences).

7.12 Software

Windows 2000, XP (Microsoft), MS Office (Microsoft), Mozilla Firefox, Adobe Acrobat 7.0, Photoshop 7.0 (Adobe), Corel Draw (Corel), TCS (Leica), Reference Manager 11 (Thompson ISI research soft), Image Reader LAS-3000 (Fuji Film Science), SDS 2.2.2 (Applied Biosystems), MetaMorph 5.8 (Molecular Analytical Technologies)

8 Methods

8.1 Cell culture methods

8.1.1 Passaging of cells

Adherent cells were grown in 75 cm^2-flasks at 37°C in a water-saturated, 5% CO_2-containing atmosphere. Medium of sub-confluent cells was aspirated, cells were washed with PBS and treated with 1 ml Trypsin/EDTA at 37°C for 5-10 minutes. After cell detachment, 19 ml of fresh medium was added and a fraction was transferred to a new flask which volume depended on the desired dilution and the following passaging date. Medium was added to reach a final volume of 12 ml and the cells were grown again under the aforementioned conditions.

In order to seed a distinct amount of cells (e.g. for experiments in multiwell plates), a small volume (30 µl) of cell suspension was transferred to a Neubauer chamber and all cells within the four big squares were counted, with one square containing a volume of 0.1 µl. To obtain the concentration of cells/ml, the following calculation was applied:

$$cells/ml = \frac{counted\ cells}{0.4\ \mu l \times 1000\ \frac{\mu l}{ml}}$$

Cells where then seeded in a volume according to the desired dilution and applied cell culture dish.

Cell culture vessel	Growth area (cm^2)	Number of cells
Multiwell-plates		
24-well	2	2.5 x 10^5
12-well	4	5 x 10^5
6-well	9,5	1 x 10^6
Dishes		
100 mm	56	7 x 10^6
Flasks		
250-300 ml	75	1 x 10^7
650-750 ml	162-175	2 x 10^7

8.1.2 Transfection of cells

8.1.2.1 Transfection of plasmid DNA

Adherent cells were detached (see Section 8.1.1) and seeded in the desired dilution and vessel. The volume was adjusted to 250 µl if the transfection was carried out in a 24-well plate. For each well, 0.35 µl of Lipofectamine (Invitrogen) and 3.5 µg of the DNA construct were diluted separately in 50 µl of Optimem. The solutions were incubated at RT for 5 min, mixed and incubated again for 25 min to allow the formation of DNA-lipid complexes. The mixture was then poured dropwise onto the cells and transfection was performed for 4-5h. Transfection solution was aspirated and the cells were supplied with fresh medium. To obtain proper protein expression and localization, cells were grown o/n under growth conditions.

8.1.3 Infection of cells with *N. gonorrhoeae*

Bacteria were plated on GC agar plates two days befor infection and grown in a water-saturated, 5% CO_2-containing atmosphere at 37°C. The day after, colonies were checked under a binocular eyepiece for pilus and Opa expression status by colony morphology (Swanson, 1978). 5 clones with the desired properties were picked and streaked onto a new plate. The plate(s) were then grown over night and infection of cells was carried out at least 16 hours after streaking to avoid autolysis of bacteria.

One hour prior to infection, culture medium was removed from cells, cells were washed once with medium (w/o FCS) and fresh medium (w/o FCS) was applied to the cells. Cells were allowed to rest for 1-2 hours. Using a cotton bud, a swab of bacteria was then taken from a GC agar plate and resuspended in 5 ml of cell culture medium by short vortexing and bacterial concentration was assessed by measuring the light absorption of a 1:10 dilution of the suspension at 550 nm in a photometer. In order to obtain the desired multiplicity of infection (MOI), the calculated volume of the bacterial suspension was added to the cells. For time course experiments, bacteria where centrifuged onto the cells for 3 minutes at 120 g at room temperature. Infected cells were subsequently placed back in water-saturated, 5% CO_2-containing atmosphere at 37°C.

8.1.4 Gentamicin protection assay

A standard procedure for the investigation of invasion or internalization of pathogenic bacteria by either professional phagocytic cells like macrophages or non-phagocytic, e.g. epithelial cells, is the gentamicin protection assay. The herein applied antibiotic gentamicin cannot overcome lipid bilayers, with the consequence that bacteria which adhere to the cell surface are killed whereas internalized bacteria are protected and survive the treatment. Although there is some discrepancy in sensitivity between results drawn from immuno-fluorescence microscopy and gentamicin assays, it provides comparable data of different infection conditions of a certain host-pathogen pair. To asses the ratio of internalized to infecting (i.e. adherent plus internalized) bacteria, half of the cell samples is subjected only to lysis without preceding gentamicin treatment.

Cells were seeded in 24-well plates 40-48 hours prior to infection. 90 minutes before addition of bacteria, cells were washed with RPMI 1640 (w/o FCS) and fresh medium (w/o FCS) was added to the wells. Cells were infected for two hours and washed hereafter three times using RPMI 1640 medium and 10 ml plastic pipettes to remove non-adherent bacteria. One half of infected cells was treated with 100 µl of a sterile 1% saponin solution preheated to 37°C for 7-10 minutes and cells were lysed by vigorous pipetting. 900 µl of RPMI were added to each well and serial dilutions resulting in 10^{-1}- to 10^{-4}-dilutions were produced. From each of the 10^{-2}- to 10^{-4}-dilutions, 25 µl were plated on GC agar plates and bacteria were allowed to grow under the above mentioned conditions. The other half was incubated with a 50 µM gentamicin/RPMI 1640 solution and incubated for 2 hours under growth conditions. Thereafter, cells were washed three times and subjected to saponin lysis as described above. 25 µl of the undiluted, 10^{-1}- and 10^{-2}-diluted samples were plated onto GC agar plates and incubated under growth conditions. The next day, colonies were counted and the ratio of corresponding samples was calculated to obtain the invasion rate (internalized bacteria vs. infecting bacteria).

8.2 Growth and manipulation of bacteria

8.2.1 Growth of *N. gonorrhoeae*

Bacteria were grown on GC agar plates in a water-saturated, 5% CO_2-containing atmosphere over night. The next day, bacteria were looked over for expression of

Opa outer membrane proteins and pili by assesing colony morphology (Swanson, 1978) under a binocular eyepiece. In the case of strain N280, colonies with a P^+, Opa^--appearance were selected whereas P^-, Opa^+- colonies were picked when strain N313 was checked. Bacteria were streaked on new plates and grown over night, not exceeding 16 hours to prevent bacterial autolysis. Bacteria were then resuspended in skim milk by vigorous shaking and transferred to a cryo-tube filled with glass perls for the generation of stock solutions. Suspension was aspirated immediately and thoroughly to obtain fluid-coated perls. The tube was stored at -80°C.

8.2.2 Growth of *E. coli*

E. coli were grown either in LB broth for plasmid DNA purification or stock solution generation or on LB agar plates for colony selection or for the plating of transformation suspensions. For the growth in liquids supplemented with the appropriate antibiotic, vessels were shaken at 220 rounds per minute (rpm) at 37°C, bacteria on plates were grown under *N. gonorrhoeae* growth conditions (see Section 8.2.1). For the preparation of stock solutions, 300 µl of glycerol (86%) was added to 700 µl of an overnight culture and the suspension was stored at -80°C.

8.2.3 Preparation of DNA competent *E. coli*

A culture of 100 ml LB broth was inoculated with *E. coli* (strain DH5α) and grown at 37°C and shaking at 220 rpm until an OD600 of 0.4-0.5 was reached. After cooling the culture for 30 minutes on ice, bacteria were centrifuged at 5000 rpm for 10 minutes at 4°C (Sorvall, SLA-1500 rotor). The pellet was resuspended in 50 ml of 0.1 M $CaCl_2$-solution and kept on ice for 30 minutes. After a second centrifugation with the same settings, the resulting pellet was resuspended in 5 ml of a 0.1 M $CaCl_2$/10% (v/v) glycerol solution and aliquots of 100 µl were shock-frozen in liquid nitrogen and stored at -80°C.

8.2.4 Transformation of *E. coli*

1 ng of plasmid DNA was gently mixed with a 50 µl suspension containing DNA competent bacteria in an 1.5 ml reaction tube and incubated for 30 minutes on ice. Next, the tube was placed in a waterbath or heat block adjusted to 42°C, incubated

for 90 seconds and immediately transferred to ice where it was left for 1-2 minutes. A volume of 800 µl of SOC medium preheated to 37°C was added to the bacteria and the suspension was shaken at 220 rpm at 37°C for one hour. After the regeneration step, 150 µl were plated on agar plates containing the appropriate antibiotic and bacteria were grown on the inverted plates at 37°C over night.

8.3 Nucleic acid methods

8.3.1 Preparative isolation of plasmid DNA

Plasmid DNA was purified with either the NucleoBond AX Plasmid DNA Purification Kit by Macherey-Nagel or the Qiafilter^{TM} Plasmid Purification kit. Procedure described here applies for the latter. 5ml of LB broth were supplemented with the required antibiotic and inoculated with *E. coli* carrying the plasmid DNA to be purified. The culture was incubated under shaking at 220 rpm at 37°C for 6-8 h and 500 µl served as inoculant for 100 ml LB broth (plus antibiotic) in a flask which was incubated o/n under the same conditions. Bacteria were harvested by centrifugation at 6000x g for 15 min at 4°C (Sorvall, SLA-1500 rotor) and resuspended in 4 ml Rnase A-containing buffer P1. 4 ml of lysis buffer P2 was added, tube was inverted vigorously 4-6 times and suspension was incubated for 5 min maximum at RT. For neutralization, 4ml of chilled buffer P3 were added. The lysate was then poured into a QIAfilter cartridge and incubated for 10 min at room temperature. A QIAGEN-tip 100 column was equilibrated with 4 ml of buffer QBT and the lysate was extruded through the cartridge filter by inserting the plunger into the affinity column. The column was washed twice with 10 ml of buffer QC and plasmid DNA was eluted with 5 ml of buffer QF into a centrifugation tube. Precipitation of DNA was obtained by addition of 3.5 ml isopropanol and centrifugation at 15 000x g for 30 min at 4°C (SS34 rotor, Sorvall). The pellet was washed with 2 ml of 70% ethanol, centrifuged again at 15 000x g for 10 min and, after drying, resuspended in 300 µl of TE buffer or ddH$_2$O.

DNA concentration was determined photometricaly by diluting plasmid DNA solution 1:100 and measuring the absorption at a wavelength of 260 nm in a glass cuvette with a thickness of 1 cm. Under these conditions, the DNA concentration is 50 µg/ml if ΔE_{260}=1. The ratio of the 260/280 values should equal 1.8, values lower than 1.6 or higher than 2.0 indicate protein contamination.

8.3.2 RNA purification from adherent cells

For RNA purification, the Rneasy® kit from Qiagen was applied. Cells were cultured in 6-well plates, washed twice with PBS and lysed with 350 µl of RLT buffer. 350 µl of 70% ethanol were added and the lysate was transferred to an Rneasy spin column placed in a collection tube. The column was centrifuged for 15 sec at 8000x g at RT and the flow-through was discarded. 700 µl of buffer RW1 were added to the column which was centrifuged again for 15 sec at 8000x g. After discarding the flow-through, the column membrane was washed twice with 500 µl of buffer RPE by centrifugation at 8000x g for 15 sec in the first and for 2 min in the second step to remove ethanol. The column was then placed in a new tube and centrifuged for 1 min at full speed to eliminate possible carryover of RPE. The column was now transferred to a 1.5 ml collection tube centrifuged for 1 min at 8000g after the addition of 30-50 µl of Rnase-free water to elute the RNA.

8.3.3 Quantitative real-time polymerase chain reaction (qRT-PCR)

The principle of qRT-PCR is based on the detection and quantification of fluorescent reporter compounds. The amount of these compounds is in proportion to the synthesized PCR product and, in parallel, to the mRNA concentration of the gene of interest. Due to the high sensitivity and accuracy of this method, even rare transcripts and small changes in the expression level can be detected.

As one of several applicable reporter compounds, SYBR® green was chosen for the experiments in this work. This fluorescent dye intercalates between the base pairs of DNA and the resulting fluorescence signal increases proportional to the amount of PCR product.

Here, qRT-PCR analyzes were carried out with the QuantiTect™ SYBR® green RT-PCR kit from Qiagen. The qRT-PCR reaction was carried out in 96-well microtiter plates in a volume of 25 µl per RNA sample comprising 10 µl of SYBR® green reaction buffer, 0.5 µl of 10 pmol/µl primer, 0.25 µl reverse transcriptase, 4.25 µl of RNAse-free water and 10 µl of 10 ng/µl RNA. To eradicate air bubbles, the plate was centrifuged for 3 min at 2500 g. The plate was placed in an Abi Prism 7900HT qRT-PCR machine (Applied Biosystems) and amplification of the complementary DNA (cDNA) was achieved in two steps: first, the RNA was reversed transcribed to DNA by incubating the samples at 50°C for 30 min. In a second step, cDNA was first

Methods

dissociated by heating up the samples to 95°C for 15 min and then amplified by 45 PCR cycles. One cycle was composed of a denaturing step at 94°C for 20 sec, followed by an annealing at 60°C for 35 sec and an elongation step at 72°C for 35 seconds.
Expression levels of examined genes were normalized with the internal standard GAP-DH (glyceraldehyde-3-dehyodrogenase). The data was analyzed using the Abi Prism 7001 software.

8.3.4 RNA interference (RNAi)

Due to the convenient appliance in molecular biology experiments, RNAi is the method of choice to specifically downregulate gene expression. It is based on the capability of RNA oligonucleotides that own a complementary sequence to the target gene to interfere with messenger RNA (mRNA) translation by either blocking the proceeding of the ribosome on the mRNA molecule or by initiating the degradation of the target mRNA. The latter occurs in eukaryotic cells by the ATP-dependent cleavage of double-stranded RNA (dsRNA) molecules by the Rnase III family member Dicer into smaller oligonucleotides termed siRNAs (small interfering RNAs). These 20-25 nucleotides comprising molecules own a phosphorylated 5'-end and a non-phosphorylated 3'-end. One siRNA molecule associates with the argonaute endonuclease to form the RNA induced silencing complex (RISC) in which the RNA strands are separated from each other. The so called guide strand remains in the complex and targets the complementary mRNA to allow its degradation by the argonaute protein (Dykxhoorn *et al.*, 2003).
Lyophilisized oligonucleotides were resuspended in siRNA Suspension Buffer (Qiagen) to obtain a 20 µM solution. The suspension was stored in 5 µl aliquots and stored at -20°C.
For siRNA transfection of adherent cells, which was performed in 24-well plates, for each well 2 µl of a 2 µM siRNA solution was diluted in 60 µl of cell culture medium (w/o FCS) and incubated at RT for 10 min. Hereafter, 2 µl of HiPerFect transfection reagent (Qiagen) were added and the mixture was incubated again for 10 min. 100 µl of FCS-containing cell culture medium was added and the transfection solution was added to the cells freshly seeded (see Section 8.1.1) in the same volume as the applied siRNA mixture. Because siRNAs differ in their knockdown capacity, the

Methods

number of seeded cells has to be titrated against a fixed volume of transfection reagent in order to obtain an efficient protein knockdown. Transfection mixture was incubated with the cells for 3 days and protein levels of cell lysates were validated by Western blot.

8.4 Protein biochemical methods

8.4.1 Discontinuous SDS polyacrylamide gel electrophoresis (SDS-PAGE)

By virtue of this technology complex mixtures of proteins (e.g. cell lysates) can be separated according to their molecular weight. Therefore, protein samples are denatured in 1x Laemmli buffer (sample buffer) by boiling at 95°C for at 5 minutes. Due to the association of sodium dodecyl sulphate (SDS) with proteins with a ratio of 1.4 g SDS/g protein, the peptides are not only denatured but carry also a strong negative charge, warranting the separation by size. In the case of cell lysates, samples were boiled for 20 minutes to degrade DNA, a measure that facilitates the gel loading procedure.

For the production of polyacrylamid gels, separating gel solution was poured between two glass plates of the mini-PROTEAN® II Electrophoresis Cell system to obtain a gel of the size 5.5 x 8.5 cm^2. Different pore sizes of the gel were established by varying the acrylamid percentage in the solution from 10-15%. The solution was covered with water to allow polymerization. After complete polymerization, water was removed and stacking gel solution was poured on top of the polymerized separating gel. Loading slots for protein samples were obtained by inserting a gel comb prior to polymerization. The comb was removed after polymerization of the stacking gel and the gel was placed in an electrophoresis tank which was then filled with running buffer. Protein samples where loaded onto the gel with a Hamilton Microliter™ syringe. To start electrophoresis, a voltage of 90 volt was applied and raised to 160 volt after the samples had passed the stacking gel. Proteins were allowed to migrate at least until the elution of bromphenol blue. For better separation, migration was extended and monitored by the disappearance of NEB® prestained protein marker bands. After protein separation, the gel was detached from glass plates and the stacking gel was cut from the separation gel and discarded. The gel was then either processed in immuno blotting or stained with Coomassie®.

8.4.2 Coomassie® stain of protein gels

Staining with Coomassie is a standard method for the staining of proteins in SDS-PAGE gels. Protein gels are therefore incubated in Coomassie® brilliant blue R-250 solution for ½ to 1h under gentle shaking, rinsed with water and destainend in destain solution for 1h – o/n. Gels were either sealed in a moistered plastic envelope for further use or dried on Whatman paper at 70°C for 2h in a vacuum drier.

8.4.3 Immuno blot (Western blot)

Gel staining methods like Coomassie®- or silver nitrate-staining have the disadvantage that they are not specific for one particular protein. In addition, small amounts of proteins are not detected due to sensitivity limits. Immunoblotting overcomes both limits by the application of antibodies specific for a single polypeptide. Therefore, the antibody solution is incubated with a membrane to which the proteins have been transferred electrophoretically (Towbin et al., 1979). In order to enhance the signal, horse radish peroxidase-conjugated secondary antibodies which recognize the F_c portion of the first antibody bound to the protein are applied. The edition of a peroxidase substrate solution leads to a chemoluminescence signal which is detectable by photo films.

Proteins are transferred to a PVDF membrane either by the semi-dry or wet blot technique. In both cases, PVDF membrane with a size corresponding to the applied gel is activated by incubation in methanol for 15 sec., washed in water for 3 min and equilibrated in blotting buffer for 10 min. For semi-dry blotting, two layers of wetted Whatman® paper are placed on the platin-covered anode of the blotting chamber (Biometra), followed by the membrane, the gel and again two layers of Whatman® paper. The chamber is closed with the anode lid and a current of 50 mA/blot was applied for 1h. If the proteins were transferred by wet blot, blot column was assembled on the black part of the gel holder cassette, with one fiber pad on each side. Cassettes were closed and inserted into the electrode assembly, which was then placed in the buffer tank. A current of 250 mA was applied for a 3h transfer. For o/n transfer, current was adjusted to 150 mA. Successful transfer of proteins was checked by incubating the membrane with Ponceau S solution for 2 min and subsequent washing with dH_2O.

Unspecific antibody binding was blocked by incubating the membrane for 1 h at room temperature with a solution of 10% nonfat milk and 3% BSA in TBS-T. Afterwards, membrane was washed 5 times for 5 min with TBS-T under gentle shaking and incubated with the primary antibody solution in a 50 ml Falcon tube for 1h. The membrane was washed again 4 times for 5 min and the secondary antibody solution was applied for 50 min Solution was discarded and the membrane was subjected to a final washing step (3x, 5 min) before visualization by means of enhanced chemoluminescence (ECL) reaction.

8.4.4 Membrane stripping

Antibodies were removed from the membrane to allow the detection of a second protein which has similar migration properties compared to the previous detected one. The membrane was therefore incubated with stripping solution for 20 min at 50°C and washed with TBS-T o/n after three washing steps for 10 min to ensure complete removal of β-mercaptoethanol and SDS which interfere with antibody binding. Protein detection was then carried out as described (see Section 8.4.3)

8.4.5 Protein co-immuno precipitation

A number of 1.2×10^6 cells were seeded in a 100 mm cell culture dish to obtain a subconfluent cell layer 48h later. Prior to infection, culture medium was removed, cells were washed once with RPMI 1640 (w/o FCS) and were allowed to rest for 1h in serum-free RPMI. Cells which were either infected with *N. gonorrhoeae* or remained uninfected were washed with cold (4°C) PBS and 800 µl of chilled cell lysis buffer supplemented with Complete protease inhibitors and 1 mM Na_3PO_4 (final concentration) were added. Cells were scraped from the dish with a cell scraper and collected in a 1.5 ml reaction tube. The cell lysate was vortexed for 30 sec., incubated on ice for 10 min and centrifuged at 14 000x g for 10 min at 4°C. The supernatant was transferred in a new tube, 2 µg of antibody were added to each tube and samples were rotated o/n at 4°C. The next day, 100 µl of ImmunoPure® Immobilized Protein A/G slurry was added to the samples and rotated for another 2 h at RT to allow binding of antibody to the protein A/G beads. Affinity of antibodies to either protein A or G depends on the species in which the antibody is raised and on the IgG isotype. These considerations are not necessary if a mixture of protein A and

G is applied as in the described case. Beads are washed 4 times with 1 ml of cell lysis buffer and gathered each time by centrifugation at 2500x g for 2-3 min. After a final wash with 1 ml ddH$_2$O, remaining fluid was aspirated with a Hamilton syringe and beads were boiled at 95°C in sample buffer for 5 min. Precipitated proteins were detected by Western blotting.

8.4.6 Purification of PilC

N. gonorrhoeae MS11 strain 560 (Scheuerpflug *et al.*, 1999) was plated on 5-10 GC Agar plates and grown over night under *Neisseria* spp. growth conditions. Bacteria were resuspended in sterile filtrated PP medium supplemented with vitamin mixture, and 50 µl of the suspension were distributed with a cotton bud on each GC agar plate prepared from 8 litres of GC agar supplemented with 10 mM MgCl$_2$, 1% vitamins, 5 µg/ml tetracycline and 0.5 mM IPTG (~350 plates). Bacteria were grown over night, harvested the next day with a cell scraper and suspended in 2x 20-25 ml of buffer P1 on ice. Bacteria were lysed by sonification (4x 30 sec. with a 1 min break between each step; Cycle: 50, Output: 9) on ice. Membranes were isolated by centrifuging the suspension at 6000 rpm for 15 min at 4°C (Sorvall S S34). Supernatant containing membranes was transferred to a new tube and centrifuged again at 20 000 rpm for 1h at 4°C (Sorvall SS34). The pellet was then resuspended in buffer P2 (LDAO extraction buffer, 2 ml P2 per tube). The pellet was allowed to dissolve completely by incubation at 37°C in a water bath for 1h. Supernatant from the last centrifugation (20 000 rpm, 25°C, RT) was loaded on four columns (Pasteur pipettes plugged with glass wool and filled with 1 ml of Ni-NTA-agarose slurry from Qiagen) equilibrated with 2 ml of buffer P2. Columns were washed with 5-6 ml of buffer P3 and subsequently with 5-6 ml of buffer P4. Protein elution was achieved with 2x 0.5 ml and 1x 1 ml of imidazole-containing elution buffer P5. PilC protein solution was eluted into 45 µl of 1M Na$_2$HPO$_4$/0.5 ml of eluate for neutralization. The solution was shock-frozen in liquid nitrogen in 100 µl aliquots (0.5 ml reaction tubes). Complete elution was evoked by applying 1 ml of buffer P6 to check elution efficiency. Fractions were checked for PilC protein on an SDS-PAGE gel (20 µl/fraction).

8.5 Microscopy

8.5.1 Confocal laser scanning microscopy

8.5.1.1 Indirect immunofluorescence staining

Cells grown on 12 mm diameter cover slips in a 24-well multiwell plate were washed with PBS and fixed with 4% PFA for 30 min at RT. Cells were washed with PBS twice and autofluorescence was quenched with 0.1% glycine solution for 15 min at room temperature. Afterwards, cells were washed again with PBS and either stored for later progression at 4°C or stained immediately. Therefore, cells were permeabilized with 0.1% TritonX-100/PBS for 20 min at RT, washed twice with PBS for 5 min and blocked with 1% BSA in 0.05% Tween20/PBS for 20 min. Cover slips were then inverted and put on a 50 µl drop of primary antibody solution (1:100 in blocking solution) placed on a multiwell plate lid covered with parafilm. The lid was transferred to a moistured chamber and antibody was allowed to bind for 1h. After binding, cover slips were retransferred to the multiwell plate and washed three times with PBS. Secondary antibody binding was performed according to primary antibody binding procedure, but with a 1:150 dilution and after centrifugation of the antibody stock solution for 2 min at 10 000 *g* to clear dissociated fluorescent dye. For actin staining, phaloidin coupled with either Alexa Fluor 546 or 647 were added to the secondary antibody solution in a 1:100 dilution.

After washing for three times with PBS, cover slips were rinsed with ddH_2O and mounted on a glass slide with Mowiol.

8.5.1.2 Differential indirect immunofluorescence staining of bacteria

This staining method is applied to distinguish between extra- and intracellular bacteria. For this purpose, cells are blocked with 1% BSA in 0.05% Tween20/PBS and an additional staining procedure is carried out before permeabilization with a first antibody recognizing the bacterium of interest (here: *N. gonorrhoeae*). This results in an exclusive staining of extracellular bacteria. In a second staining step of bacteria after permeabilization, the same first antibody is applied, but the second antibody has to differ in the fluorophor coupled to it (i.e. if a secondary antibody coupled to Cy2 was used in the first step, a antibody coupled to Cy3 or Cy5 has to be applied in the second step). Thus, extracellular bacteria are stained with two secondary antibodies bearing two different fluorophors while intracellular bacteria are labelled only with one fluorophor, making them distinguishable by colour channel overlay.

8.5.1.3   Confocal laser scanning microscopy

Microscopy was carried out with a Leica TCS SP microscope, equipped with an argon/krypton laser. Specimens were examined at a 630x magnification using immersion oil. Due to the transmission filter of the microscope, three different excitation wave lengths could be applied at the same time allowing the parallel observation of three differently fluorescence labelled proteins/structures.

8.5.2   Life cell imaging microscopy

The advantage of this method is the possibility to observe proteins or bacteria marked with a fluorescent compound (Alexa Fluor, GFP) in/on living cells, whereas only fixed samples can be examined by means of confocal laser scanning microscopy.

AGS- or ME180-cells were seeded on 3.5 cm^2 glass bottom cell culture dishes and transfected with either the VASP-GFP or actin-GFP construct. The next day, cells were infected with *N. gonorrhoeae* and infection was monitored with a VT- Infinity-System (Visitron Systems) consisting of an Olympus IX81 microscope (Olympus), a VT-Infinity Galvo Scanner Confocal Head (Visitron Systems) and a Hamamatsu C9100-02 CCD camera (Hamamatsu Photonics K.K). The microscope was also equipped with an acrylic glass chamber heated to 37°C and a CO_2 enrichment accessory (Solent Scientific) to sustain the viability of cells and bacteria. Phase contrast pictures were taken using an x63 phase contrast objective (NA 1.25 oil, Olympus) and a high speed closing system. Fluorescence pictures were taken by applying an excitation wave length of 488 nm (100 mV) and 568 nm (50 mV) and a 488bp/568bp/647bp emission filter set (Chroma Technologies). Laser light was generated with an Innova® 70C Spectrum device (Coherent). Pictures were acquired every 20 sec over a period of up to 4h and analyzed with the acquisition programme Metamorph 5.8 (Molecular Analytical Technologies).

9 References

1. AbdelRahman, Y.M. and Belland, R.J. (2005). The chlamydial developmental cycle. *FEMS Microbiol.Rev.* **29** (5): 949-959.

2. Allan, I. and Pearce, J.H. (1987). Association of Chlamydia trachomatis with mammalian and cultured insect cells lacking putative chlamydial receptors. *Microb.Pathog.* **2** (1): 63-70.

3. Anderson, H.A., Chen, Y., and Norkin, L.C. (1996). Bound simian virus 40 translocates to caveolin-enriched membrane domains, and its entry is inhibited by drugs that selectively disrupt caveolae. *Mol.Biol.Cell* **7** (11): 1825-1834.

4. Apicella, M.A., Ketterer, M., Lee, F.K., Zhou, D., Rice, P.A., and Blake, M.S. (1996). The pathogenesis of gonococcal urethritis in men: confocal and immunoelectron microscopic analysis of urethral exudates from men infected with Neisseria gonorrhoeae. *J.Infect.Dis.* **173** (3): 636-646.

5. Ayala, B.P., Vasquez, B., Clary, S., Tainer, J.A., Rodland, K., and So, M. (2001). The pilus-induced Ca2+ flux triggers lysosome exocytosis and increases the amount of Lamp1 accessible to Neisseria IgA1 protease. *Cell Microbiol.* **3** (4): 265-275.

6. Ayala, P., Vasquez, B., Wetzler, L., and So, M. (2002). Neisseria gonorrhoeae porin P1.B induces endosome exocytosis and a redistribution of Lamp1 to the plasma membrane. *Infect.Immun.* **70** (11): 5965-5971.

7. Ayala, P., Wilbur, J.S., Wetzler, L.M., Tainer, J.A., Snyder, A., and So, M. (2005). The pilus and porin of Neisseria gonorrhoeae cooperatively induce Ca(2+) transients in infected epithelial cells. *Cell Microbiol.* **7** (12): 1736-1748.

8. Bai, L., Schuller, S., Whale, A., Mousnier, A., Marches, O., Wang, L., Ooka, T., Heuschkel, R., Torrente, F., Kaper, J.B., Gomes, T.A., Xu, J., Phillips, A.D., and Frankel, G. (2008). Enteropathogenic Escherichia coli O125:H6 triggers attaching and effacing lesions on human intestinal biopsy specimens independently of Nck and TccP/TccP2. *Infect.Immun.* **76** (1): 361-368.

9. Bakre, M.M., Zhu, Y., Yin, H., Burton, D.W., Terkeltaub, R., Deftos, L.J., and Varner, J.A. (2002). Parathyroid hormone-related peptide is a naturally occurring, protein kinase A-dependent angiogenesis inhibitor. *Nat.Med.* **8** (9): 995-1003.

10. Baorto, D.M., Gao, Z., Malaviya, R., Dustin, M.L., van der, M.A., Lublin, D.M., and Abraham, S.N. (1997). Survival of FimH-expressing enterobacteria in macrophages relies on glycolipid traffic. *Nature* **389** (6651): 636-639.

11. Barlow, D. and Phillips, I. (1978). Gonorrhoea in women. Diagnostic, clinical, and laboratory aspects. *Lancet* **1** (8067): 761-764.

12. Beyaert, R., Heyninck, K., and Van Huffel, S. (2000). A20 and A20-binding proteins as cellular inhibitors of nuclear factor-kappa B-dependent gene expression and apoptosis. *Biochem.Pharmacol.* **60** (8): 1143-1151.

13. Biais, N., Ladoux, B., Higashi, D., So, M., and Sheetz, M. (2008). Cooperative retraction of bundled type IV pili enables nanonewton force generation. *PLoS.Biol.* **6** (4): e87-

14. Bille, E., Zahar, J.R., Perrin, A., Morelle, S., Kriz, P., Jolley, K.A., Maiden, M.C., Dervin, C., Nassif, X., and Tinsley, C.R. (2005). A chromosomally integrated bacteriophage in invasive meningococci. *J.Exp.Med.* **201** (12): 1905-1913.

15. Billker, O., Popp, A., Brinkmann, V., Wenig, G., Schneider, J., Caron, E., and Meyer, T.F. (2002). Distinct mechanisms of internalization of Neisseria gonorhoeae by members of

the CEACAM receptor family involving Rac1- and Cdc42-dependent and -independent pathways. *EMBO J.* **21** (4): 560-571.

16. Boettcher, J., Kirchner, M., Churin, Y., Thorn, H., Brinkmann, V., and Meyer, T. (2008). Type IV pili producing bacteria recruit caveolin to escape premature host cell uptake. *in preparation*

17. Booy, R. and Kroll, J.S. (1998). Bacterial meningitis and meningococcal infection. *Curr.Opin.Pediatr.* **10** (1): 13-18.

18. Breau, W.C., Atwood, W.J., and Norkin, L.C. (1992). Class I major histocompatibility proteins are an essential component of the simian virus 40 receptor. *J.Virol.* **66** (4): 2037-2045.

19. Brindle, N.P., Holt, M.R., Davies, J.E., Price, C.J., and Critchley, D.R. (1996). The focal-adhesion vasodilator-stimulated phosphoprotein (VASP) binds to the proline-rich domain in vinculin. *Biochem.J.* **318 (Pt 3)** 753-757.

20. Brown, D.A. and London, E. (1998). Structure and origin of ordered lipid domains in biological membranes. *J.Membr.Biol.* **164** (2): 103-114.

21. Brown, D.A. and Rose, J.K. (1992). Sorting of GPI-anchored proteins to glycolipid-enriched membrane subdomains during transport to the apical cell surface. *Cell* **68** (3): 533-544.

22. Brown, M.T. and Cooper, J.A. (1996). Regulation, substrates and functions of src. *Biochim.Biophys.Acta* **1287** (2-3): 121-149.

23. Burch, C.L., Danaher, R.J., and Stein, D.C. (1997). Antigenic variation in Neisseria gonorrhoeae: production of multiple lipooligosaccharides. *J.Bacteriol.* **179** (3): 982-986.

24. Campellone, K.G., Robbins, D., and Leong, J.M. (2004). EspFU is a translocated EHEC effector that interacts with Tir and N-WASP and promotes Nck-independent actin assembly. *Dev.Cell* **7** (2): 217-228.

25. Capecchi, B., Adu-Bobie, J., Di Marcello, F., Ciucchi, L., Masignani, V., Taddei, A., Rappuoli, R., Pizza, M., and Arico, B. (2005). Neisseria meningitidis NadA is a new invasin which promotes bacterial adhesion to and penetration into human epithelial cells. *Mol.Microbiol.* **55** (3): 687-698.

26. Casey, S.G., Shafer, W.M., and Spitznagel, J.K. (1986). Neisseria gonorrhoeae survive intraleukocytic oxygen-independent antimicrobial capacities of anaerobic and aerobic granulocytes in the presence of pyocin lethal for extracellular gonococci. *Infect.Immun.* **52** (2): 384-389.

27. Caugant, D.A., Hoiby, E.A., Magnus, P., Scheel, O., Hoel, T., Bjune, G., Wedege, E., Eng, J., and Froholm, L.O. (1994). Asymptomatic carriage of Neisseria meningitidis in a randomly sampled population. *J.Clin.Microbiol.* **32** (2): 323-330.

28. Chakraborty, T., Ebel, F., Domann, E., Niebuhr, K., Gerstel, B., Pistor, S., Temm-Grove, C.J., Jockusch, B.M., Reinhard, M., Walter, U., and . (1995). A focal adhesion factor directly linking intracellularly motile Listeria monocytogenes and Listeria ivanovii to the actin-based cytoskeleton of mammalian cells. *EMBO J.* **14** (7): 1314-1321.

29. Chen, H.D. and Frankel, G. (2005). Enteropathogenic Escherichia coli: unravelling pathogenesis. *FEMS Microbiol.Rev.* **29** (1): 83-98.

30. Claus, H., Maiden, M.C., Wilson, D.J., McCarthy, N.D., Jolley, K.A., Urwin, R., Hessler, F., Frosch, M., and Vogel, U. (2005). Genetic analysis of meningococci carried by children and young adults. *J.Infect.Dis.* **191** (8): 1263-1271.

31. Cohen, A.W., Razani, B., Wang, X.B., Combs, T.P., Williams, T.M., Scherer, P.E., and Lisanti, M.P. (2003). Caveolin-1-deficient mice show insulin resistance and defective insulin

receptor protein expression in adipose tissue. *Am.J.Physiol Cell Physiol* **285** (1): C222-C235.

32. Conner, S.D. and Schmid, S.L. (2003). Regulated portals of entry into the cell. *Nature* **422** (6927): 37-44.

33. Couet, J., Li, S., Okamoto, T., Ikezu, T., and Lisanti, M.P. (1997a). Identification of peptide and protein ligands for the caveolin-scaffolding domain. Implications for the interaction of caveolin with caveolae-associated proteins. *J.Biol.Chem.* **272** (10): 6525-6533.

34. Couet, J., Sargiacomo, M., and Lisanti, M.P. (1997b). Interaction of a receptor tyrosine kinase, EGF-R, with caveolins. Caveolin binding negatively regulates tyrosine and serine/threonine kinase activities. *J.Biol.Chem.* **272** (48): 30429-30438.

35. Criss, A.K. and Seifert, H.S. (2006). Gonococci exit apically and basally from polarized epithelial cells and exhibit dynamic changes in type IV pili. *Cell Microbiol.* **8** (9): 1430-1443.

36. Damm, E.M., Pelkmans, L., Kartenbeck, J., Mezzacasa, A., Kurzchalia, T., and Helenius, A. (2005). Clathrin- and caveolin-1-independent endocytosis: entry of simian virus 40 into cells devoid of caveolae. *J.Cell Biol.* **168** (3): 477-488.

37. Delpiano, M.A. and Acker, H. (1991). Hypoxia increases the cyclic AMP content of the cat carotid body in vitro. *J.Neurochem.* **57** (1): 291-297.

38. Dietzen, D.J., Hastings, W.R., and Lublin, D.M. (1995). Caveolin is palmitoylated on multiple cysteine residues. Palmitoylation is not necessary for localization of caveolin to caveolae. *J.Biol.Chem.* **270** (12): 6838-6842.

39. Dong, J.M., Leung, T., Manser, E., and Lim, L. (1998). cAMP-induced morphological changes are counteracted by the activated RhoA small GTPase and the Rho kinase ROKalpha. *J.Biol.Chem.* **273** (35): 22554-22562.

40. Dransfield, D.T., Bradford, A.J., Smith, J., Martin, M., Roy, C., Mangeat, P.H., and Goldenring, J.R. (1997). Ezrin is a cyclic AMP-dependent protein kinase anchoring protein. *EMBO J.* **16** (1): 35-43.

41. Du, Y. and Arvidson, C.G. (2006). RpoH mediates the expression of some, but not all, genes induced in Neisseria gonorrhoeae adherent to epithelial cells. *Infect.Immun.* **74** (5): 2767-2776.

42. Duncan, M.J., Li, G., Shin, J.S., Carson, J.L., and Abraham, S.N. (2004). Bacterial penetration of bladder epithelium through lipid rafts. *J.Biol.Chem.* **279** (18): 18944-18951.

43. Dykxhoorn, D.M., Novina, C.D., and Sharp, P.A. (2003). Killing the messenger: short RNAs that silence gene expression. *Nat.Rev.Mol.Cell Biol.* **4** (6): 457-467.

44. Edwards, J.L. and Apicella, M.A. (2005). I-domain-containing integrins serve as pilus receptors for Neisseria gonorrhoeae adherence to human epithelial cells. *Cell Microbiol.* **7** (8): 1197-1211.

45. Edwards, J.L., Brown, E.J., Uk-Nham, S., Cannon, J.G., Blake, M.S., and Apicella, M.A. (2002). A co-operative interaction between Neisseria gonorrhoeae and complement receptor 3 mediates infection of primary cervical epithelial cells. *Cell Microbiol.* **4** (9): 571-584.

46. Ellerbroek, S.M., Wennerberg, K., and Burridge, K. (2003). Serine phosphorylation negatively regulates RhoA in vivo. *J.Biol.Chem.* **278** (21): 19023-19031.

47. Eugene, E., Hoffmann, I., Pujol, C., Couraud, P.O., Bourdoulous, S., and Nassif, X. (2002). Microvilli-like structures are associated with the internalization of virulent capsulated Neisseria meningitidis into vascular endothelial cells. *J.Cell Sci.* **115** (Pt 6): 1231-1241.

48. Feoktistov, I., Goldstein, A.E., and Biaggioni, I. (2000). Cyclic AMP and protein kinase A stimulate Cdc42: role of A(2) adenosine receptors in human mast cells. *Mol.Pharmacol.* **58** (5): 903-910.

49. Frasch, C.E., Zollinger, W.D., and Poolman, J.T. (1985). Serotype antigens of Neisseria meningitidis and a proposed scheme for designation of serotypes. *Rev.Infect.Dis.* **7** (4): 504-510.

50. Freitag, N.E., Seifert, H.S., and Koomey, M. (1995). Characterization of the pilF-pilD pilus-assembly locus of Neisseria gonorrhoeae. *Mol.Microbiol.* **16** (3): 575-586.

51. Fujimoto, T., Kogo, H., Nomura, R., and Une, T. (2000). Isoforms of caveolin-1 and caveolar structure. *J.Cell Sci.* **113 Pt 19** 3509-3517.

52. Gertler, F.B., Niebuhr, K., Reinhard, M., Wehland, J., and Soriano, P. (1996). Mena, a relative of VASP and Drosophila Enabled, is implicated in the control of microfilament dynamics. *Cell* **87** (2): 227-239.

53. Gibbs, C.P., Reimann, B.Y., Schultz, E., Kaufmann, A., Haas, R., and Meyer, T.F. (1989). Reassortment of pilin genes in Neisseria gonorrhoeae occurs by two distinct mechanisms. *Nature* **338** (6217): 651-652.

54. Giron, J.A., Ho, A.S., and Schoolnik, G.K. (1991). An inducible bundle-forming pilus of enteropathogenic Escherichia coli. *Science* **254** (5032): 710-713.

55. Glenney, J.R., Jr. (1989). Tyrosine phosphorylation of a 22-kDa protein is correlated with transformation by Rous sarcoma virus. *J.Biol.Chem.* **264** (34): 20163-20166.

56. Gomez-Duarte, O.G., Dehio, M., Guzman, C.A., Chhatwal, G.S., Dehio, C., and Meyer, T.F. (1997). Binding of vitronectin to opa-expressing Neisseria gonorrhoeae mediates invasion of HeLa cells. *Infect.Immun.* **65** (9): 3857-3866.

57. Goosney, D.L., DeVinney, R., Pfuetzner, R.A., Frey, E.A., Strynadka, N.C., and Finlay, B.B. (2000). Enteropathogenic E. coli translocated intimin receptor, Tir, interacts directly with alpha-actinin. *Curr.Biol.* **10** (12): 735-738.

58. Gouin, E., Gantelet, H., Egile, C., Lasa, I., Ohayon, H., Villiers, V., Gounon, P., Sansonetti, P.J., and Cossart, P. (1999). A comparative study of the actin-based motilities of the pathogenic bacteria Listeria monocytogenes, Shigella flexneri and Rickettsia conorii. *J.Cell Sci.* **112 (Pt 11)** 1697-1708.

59. Grassme, H., Gulbins, E., Brenner, B., Ferlinz, K., Sandhoff, K., Harzer, K., Lang, F., and Meyer, T.F. (1997). Acidic sphingomyelinase mediates entry of N. gonorrhoeae into nonphagocytic cells. *Cell* **91** (5): 605-615.

60. Grassme, H.U., Ireland, R.M., and van Putten, J.P. (1996). Gonococcal opacity protein promotes bacterial entry-associated rearrangements of the epithelial cell actin cytoskeleton. *Infect.Immun.* **64** (5): 1621-1630.

61. Griffiths, N.J., Bradley, C.J., Heyderman, R.S., and Virji, M. (2007). IFN-gamma amplifies NFkappaB-dependent Neisseria meningitidis invasion of epithelial cells via specific upregulation of CEA-related cell adhesion molecule 1. *Cell Microbiol.* **9** (12): 2968-2983.

62. Gronholm, M., Vossebein, L., Carlson, C.R., Kuja-Panula, J., Teesalu, T., Alfthan, K., Vaheri, A., Rauvala, H., Herberg, F.W., Tasken, K., and Carpen, O. (2003). Merlin links to the cAMP neuronal signaling pathway by anchoring the RIbeta subunit of protein kinase A. *J.Biol.Chem.* **278** (42): 41167-41172.

63. Handsfield, H.H. *Neisseria gonorrhoeae* in *Principles and practice of infectious diseases* edn. 3 (eds. Mandell,G.L., Douglas,R.G., and Bennett,J.E.) 1613-1631 (Churchill Livingstone, New York, 1990)

References

64. Haraga, A., Ohlson, M.B., and Miller, S.I. (2008). Salmonellae interplay with host cells. *Nat.Rev.Microbiol.* **6** (1): 53-66.

65. Hardt, W.D., Chen, L.M., Schuebel, K.E., Bustelo, X.R., and Galan, J.E. (1998). S. typhimurium encodes an activator of Rho GTPases that induces membrane ruffling and nuclear responses in host cells. *Cell* **93** (5): 815-826.

66. Hart, C.A. and Cuevas, L.E. (1997). Meningococcal disease in Africa. *Ann.Trop.Med.Parasitol.* **91** (7): 777-785.

67. Harvey, H.A., Jennings, M.P., Campbell, C.A., Williams, R., and Apicella, M.A. (2001). Receptor-mediated endocytosis of Neisseria gonorrhoeae into primary human urethral epithelial cells: the role of the asialoglycoprotein receptor. *Mol.Microbiol.* **42** (3): 659-672.

68. Hauck, C.R. and Meyer, T.F. (1997). The lysosomal/phagosomal membrane protein h-lamp-1 is a target of the IgA1 protease of Neisseria gonorrhoeae. *FEBS Lett.* **405** (1): 86-90.

69. Hauck, C.R., Meyer, T.F., Lang, F., and Gulbins, E. (1998). CD66-mediated phagocytosis of Opa52 Neisseria gonorrhoeae requires a Src-like tyrosine kinase- and Rac1-dependent signalling pathway. *EMBO J.* **17** (2): 443-454.

70. Heckels, J.E. (1981). Structural comparison of Neisseria gonorrhoeae outer membrane proteins. *J.Bacteriol.* **145** (2): 736-742.

71. Higashi, D.L., Lee, S.W., Snyder, A., Weyand, N.J., Bakke, A., and So, M. (2007). Dynamics of Neisseria gonorrhoeae attachment: microcolony development, cortical plaque formation, and cytoprotection. *Infect.Immun.* **75** (10): 4743-4753.

72. Hoffmann, I., Eugene, E., Nassif, X., Couraud, P.O., and Bourdoulous, S. (2001). Activation of ErbB2 receptor tyrosine kinase supports invasion of endothelial cells by Neisseria meningitidis. *J.Cell Biol.* **155** (1): 133-143.

73. Holmes, K.K., Eschenbach, D.A., and Knapp, J.S. (1980). Salpingitis: overview of etiology and epidemiology. *Am.J.Obstet.Gynecol.* **138** (7 Pt 2): 893-900.

74. Ikezu, T., Ueda, H., Trapp, B.D., Nishiyama, K., Sha, J.F., Volonte, D., Galbiati, F., Byrd, A.L., Bassell, G., Serizawa, H., Lane, W.S., Lisanti, M.P., and Okamoto, T. (1998). Affinity-purification and characterization of caveolins from the brain: differential expression of caveolin-1, -2, and -3 in brain endothelial and astroglial cell types. *Brain Res.* **804** (2): 177-192.

75. Isshiki, M. and Anderson, R.G. (2003). Function of caveolae in Ca2+ entry and Ca2+-dependent signal transduction. *Traffic.* **4** (11): 717-723.

76. Jay, D., Garcia, E.J., Lara, J.E., Medina, M.A., and de, I.L., I (2000). Determination of a cAMP-dependent protein kinase phosphorylation site in the C-terminal region of human endothelial actin-binding protein. *Arch.Biochem.Biophys.* **377** (1): 80-84.

77. Jeanteur, D., Lakey, J.H., and Pattus, F. (1991). The bacterial porin superfamily: sequence alignment and structure prediction. *Mol.Microbiol.* **5** (9): 2153-2164.

78. Johnson, A.P. (1983). The pathogenic potential of commensal species of Neisseria. *J.Clin.Pathol.* **36** (2): 213-223.

79. Jonsson, A.B., Ilver, D., Falk, P., Pepose, J., and Normark, S. (1994). Sequence changes in the pilus subunit lead to tropism variation of Neisseria gonorrhoeae to human tissue. *Mol.Microbiol.* **13** (3): 403-416.

80. Kallstrom, H., Islam, M.S., Berggren, P.O., and Jonsson, A.B. (1998). Cell signaling by the type IV pili of pathogenic Neisseria. *J.Biol.Chem.* **273** (34): 21777-21782.

References

81. Kallstrom, H., Liszewski, M.K., Atkinson, J.P., and Jonsson, A.B. (1997). Membrane cofactor protein (MCP or CD46) is a cellular pilus receptor for pathogenic Neisseria. *Mol.Microbiol.* **25** (4): 639-647.

82. Kellogg, D.S., Jr., Peacock, W.L., Jr., DEACON, W.E., BROWN, L., and PIRKLE, D.I. (1963). Neisseria gonorrhoeae. I. Virulence genetically linked to clonal variation. *J.Bacteriol.* **85** 1274-1279.

83. Kenny, B., DeVinney, R., Stein, M., Reinscheid, D.J., Frey, E.A., and Finlay, B.B. (1997). Enteropathogenic E. coli (EPEC) transfers its receptor for intimate adherence into mammalian cells. *Cell* **91** (4): 511-520.

84. King, G.J. and Swanson, J. (1978). Studies on gonococcus infection. XV. Identification of surface proteins of Neisseria gonorrhoeae correlated with leukocyte association. *Infect.Immun.* **21** (2): 575-584.

85. Kirchner, M., Heuer, D., and Meyer, T.F. (2005). CD46-Independent Binding of Neisserial Type IV Pili and the Major Pilus Adhesin, PilC, to Human Epithelial Cells. *Infect.Immun.* **73** (5): 3072-3082.

86. Kleist, E. and Moi, H. (1993). Transmission of gonorrhoea through an inflatable doll. *Genitourin.Med.* **69** (4): 322-

87. Knapp, J.S. (1988). Historical perspectives and identification of Neisseria and related species. *Clin.Microbiol.Rev.* **1** (4): 415-431.

88. Krause, M., Dent, E.W., Bear, J.E., Loureiro, J.J., and Gertler, F.B. (2003). Ena/VASP proteins: regulators of the actin cytoskeleton and cell migration. *Annu.Rev.Cell Dev.Biol.* **19** 541-564.

89. Kuijpers, T.W., Hoogerwerf, M., van der Laan, L.J., Nagel, G., van der Schoot, C.E., Grunert, F., and Roos, D. (1992). CD66 nonspecific cross-reacting antigens are involved in neutrophil adherence to cytokine-activated endothelial cells. *J.Cell Biol.* **118** (2): 457-466.

90. Kupsch, E.M., Knepper, B., Kuroki, T., Heuer, I., and Meyer, T.F. (1993). Variable opacity (Opa) outer membrane proteins account for the cell tropisms displayed by Neisseria gonorrhoeae for human leukocytes and epithelial cells. *EMBO J.* **12** (2): 641-650.

91. LABREC, E.H., SCHNEIDER, H., Magnani, T.J., and FORMAL, S.B. (1964). Epitehelial cell penetration as an essential step in the pathogenesis of bacillary dysentry. *J.Bacteriol.* **88** (5): 1503-1518.

92. Lambotin, M., Hoffmann, I., Laran-Chich, M.P., Nassif, X., Couraud, P.O., and Bourdoulous, S. (2005). Invasion of endothelial cells by Neisseria meningitidis requires cortactin recruitment by a phosphoinositide-3-kinase/Rac1 signalling pathway triggered by the lipo-oligosaccharide. *J.Cell Sci.* **118** (Pt 16): 3805-3816.

93. Lang, D.M., Lommel, S., Jung, M., Ankerhold, R., Petrausch, B., Laessing, U., Wiechers, M.F., Plattner, H., and Stuermer, C.A. (1998). Identification of reggie-1 and reggie-2 as plasmamembrane-associated proteins which cocluster with activated GPI-anchored cell adhesion molecules in non-caveolar micropatches in neurons. *J.Neurobiol.* **37** (4): 502-523.

94. Laurent, V., Loisel, T.P., Harbeck, B., Wehman, A., Grobe, L., Jockusch, B.M., Wehland, J., Gertler, F.B., and Carlier, M.F. (1999). Role of proteins of the Ena/VASP family in actin-based motility of Listeria monocytogenes. *J.Cell Biol.* **144** (6): 1245-1258.

95. Le, P.U., Guay, G., Altschuler, Y., and Nabi, I.R. (2002). Caveolin-1 is a negative regulator of caveolae-mediated endocytosis to the endoplasmic reticulum. *J.Biol.Chem.* **277** (5): 3371-3379.

References

96. Leemhuis, J., Boutillier, S., Schmidt, G., and Meyer, D.K. (2002). The protein kinase A inhibitor H89 acts on cell morphology by inhibiting Rho kinase. *J.Pharmacol.Exp.Ther.* **300** (3): 1000-1007.

97. Li, S., Seitz, R., and Lisanti, M.P. (1996). Phosphorylation of caveolin by src tyrosine kinases. The alpha-isoform of caveolin is selectively phosphorylated by v-Src in vivo. *J.Biol.Chem.* **271** (7): 3863-3868.

98. Lin, L., Ayala, P., Larson, J., Mulks, M., Fukuda, M., Carlson, S.R., Enns, C., and So, M. (1997). The Neisseria type 2 IgA1 protease cleaves LAMP1 and promotes survival of bacteria within epithelial cells. *Mol.Microbiol.* **24** (5): 1083-1094.

99. Liu, J., Oh, P., Horner, T., Rogers, R.A., and Schnitzer, J.E. (1997). Organized endothelial cell surface signal transduction in caveolae distinct from glycosylphosphatidylinositol-anchored protein microdomains. *J.Biol.Chem.* **272** (11): 7211-7222.

100. Lochner, A. and Moolman, J.A. (2006). The many faces of H89: a review. *Cardiovasc.Drug Rev.* **24** (3-4): 261-274.

101. Lorenzen, D.R., Dux, F., Wolk, U., Tsirpouchtsidis, A., Haas, G., and Meyer, T.F. (1999). Immunoglobulin A1 protease, an exoenzyme of pathogenic Neisseriae, is a potent inducer of proinflammatory cytokines. *J.Exp.Med.* **190** (8): 1049-1058.

102. Luan, J., Shattuck-Brandt, R., Haghnegahdar, H., Owen, J.D., Strieter, R., Burdick, M., Nirodi, C., Beauchamp, D., Johnson, K.N., and Richmond, A. (1997). Mechanism and biological significance of constitutive expression of MGSA/GRO chemokines in malignant melanoma tumor progression. *J.Leukoc.Biol.* **62** (5): 588-597.

103. Makino, S., van Putten, J.P., and Meyer, T.F. (1991). Phase variation of the opacity outer membrane protein controls invasion by Neisseria gonorrhoeae into human epithelial cells. *EMBO J.* **10** (6): 1307-1315.

104. Masi, A.T. and Eisenstein, B.I. (1981). Disseminated gonococcal infection (DGI) and gonococcal arthritis (GCA): II. Clinical manifestations, diagnosis, complications, treatment, and prevention. *Semin.Arthritis Rheum.* **10** (3): 173-197.

105. Massari, P., Ho, Y., and Wetzler, L.M. (2000). Neisseria meningitidis porin PorB interacts with mitochondria and protects cells from apoptosis. *Proc.Natl.Acad.Sci.U.S.A* **97** (16): 9070-9075.

106. Matveev, S.V. and Smart, E.J. (2002). Heterologous desensitization of EGF receptors and PDGF receptors by sequestration in caveolae. *Am.J.Physiol Cell Physiol* **282** (4): C935-C946.

107. McCaw, S.E., Schneider, J., Liao, E.H., Zimmermann, W., and Gray-Owen, S.D. (2003). Immunoreceptor tyrosine-based activation motif phosphorylation during engulfment of Neisseria gonorrhoeae by the neutrophil-restricted CEACAM3 (CD66d) receptor. *Mol.Microbiol.* **49** (3): 623-637.

108. McCormack, W.M., Stumacher, R.J., Johnson, K., and Donner, A. (1977). Clinical spectrum of gonococcal infection in women. *Lancet* **1** (8023): 1182-1185.

109. Mengaud, J., Ohayon, H., Gounon, P., Mege, R.-M., and Cossart, P. (1996). E-cadherin is the receptor for internalin, a surface protein required for entry of L. monocytogenes into epithelial cells. *Cell* **84** (6): 923-932.

110. Merrifield, C.J., Moss, S.E., Ballestrem, C., Imhof, B.A., Giese, G., Wunderlich, I., and Almers, W. (1999). Endocytic vesicles move at the tips of actin tails in cultured mast cells. *Nat.Cell Biol.* **1** (1): 72-74.

111. Merz, A.J., Enns, C.A., and So, M. (1999). Type IV pili of pathogenic Neisseriae elicit cortical plaque formation in epithelial cells. *Mol.Microbiol.* **32** (6): 1316-1332.

112. Merz, A.J., Rifenbery, D.B., Arvidson, C.G., and So, M. (1996). Traversal of a polarized epithelium by pathogenic Neisseriae: facilitation by type IV pili and maintenance of epithelial barrier function. *Mol.Med.* **2** (6): 745-754.

113. Merz, A.J. and So, M. (1997). Attachment of piliated, Opa- and Opc- gonococci and meningococci to epithelial cells elicits cortical actin rearrangements and clustering of tyrosine-phosphorylated proteins. *Infect.Immun.* **65** (10): 4341-4349.

114. Merz, A.J., So, M., and Sheetz, M.P. (2000). Pilus retraction powers bacterial twitching motility. *Nature* **407** (6800): 98-102.

115. Meyer, T.F., Frosch, M., Gibbs, C.P., Haas, R., Halter, R., Pohlner, J., and van Putten, J.P. (1988). Virulence functions and antigen variation in pathogenic Neisseriae. *Antonie Van Leeuwenhoek* **54** (5): 421-430.

116. Meyer, T.F. and van Putten, J.P. (1989). Genetic mechanisms and biological implications of phase variation in pathogenic neisseriae. *Clin.Microbiol.Rev.* **2 Suppl** S139-S145.

117. Minetti, C.A., Tai, J.Y., Blake, M.S., Pullen, J.K., Liang, S.M., and Remeta, D.P. (1997). Structural and functional characterization of a recombinant PorB class 2 protein from Neisseria meningitidis. Conformational stability and porin activity. *J.Biol.Chem.* **272** (16): 10710-10720.

118. Minshall, R.D., Tiruppathi, C., Vogel, S.M., Niles, W.D., Gilchrist, A., Hamm, H.E., and Malik, A.B. (2000). Endothelial cell-surface gp60 activates vesicle formation and trafficking via G(i)-coupled Src kinase signaling pathway. *J.Cell Biol.* **150** (5): 1057-1070.

119. Monier, S., Parton, R.G., Vogel, F., Behlke, J., Henske, A., and Kurzchalia, T.V. (1995). VIP21-caveolin, a membrane protein constituent of the caveolar coat, oligomerizes in vivo and in vitro. *Mol.Biol.Cell* **6** (7): 911-927.

120. Mora, R., Bonilha, V.L., Marmorstein, A., Scherer, P.E., Brown, D., Lisanti, M.P., and Rodriguez-Boulan, E. (1999). Caveolin-2 localizes to the golgi complex but redistributes to plasma membrane, caveolae, and rafts when co-expressed with caveolin-1. *J.Biol.Chem.* **274** (36): 25708-25717.

121. Morand, P.C., Bille, E., Morelle, S., Eugene, E., Beretti, J.L., Wolfgang, M., Meyer, T.F., Koomey, M., and Nassif, X. (2004). Type IV pilus retraction in pathogenic Neisseria is regulated by the PilC proteins. *EMBO J.* **23** (9): 2009-2017.

122. Mosleh, I.M., Boxberger, H.J., Sessler, M.J., and Meyer, T.F. (1997). Experimental infection of native human ureteral tissue with Neisseria gonorrhoeae: adhesion, invasion, intracellular fate, exocytosis, and passage through a stratified epithelium. *Infect.Immun.* **65** (8): 3391-3398.

123. Muenzner, P., Billker, O., Meyer, T.F., and Naumann, M. (2002). Nuclear factor-kappa B directs carcinoembryonic antigen-related cellular adhesion molecule 1 receptor expression in Neisseria gonorrhoeae-infected epithelial cells. *J.Biol.Chem.* **277** (9): 7438-7446.

124. Muenzner, P., Naumann, M., Meyer, T.F., and Gray-Owen, S.D. (2001). Pathogenic Neisseria trigger expression of their carcinoembryonic antigen-related cellular adhesion molecule 1 (CEACAM1; previously CD66a) receptor on primary endothelial cells by activating the immediate early response transcription factor, nuclear factor-kappaB. *J.Biol.Chem.* **276** (26): 24331-24340.

125. Muller, A., Gunther, D., Brinkmann, V., Hurwitz, R., Meyer, T.F., and Rudel, T. (2000). Targeting of the pro-apoptotic VDAC-like porin (PorB) of Neisseria gonorrhoeae to mitochondria of infected cells. *EMBO J.* **19** (20): 5332-5343.

References

126. Muller, A., Gunther, D., Dux, F., Naumann, M., Meyer, T.F., and Rudel, T. (1999). Neisserial porin (PorB) causes rapid calcium influx in target cells and induces apoptosis by the activation of cysteine proteases. *EMBO J.* **18** (2): 339-352.

127. Murray, E.G.D. and Branham, S.A. *Family VI. Neisseriaceae Prévot 1933* in *Bergey's manual of determinative bacteriology* edn. 5 (eds. Buchanan R.E.) 278-285 (The Williams & Wilkins Co., Baltimore, 1939)

128. Murray, E.G.D. and Branham, S.A. *Family VI. Neisseriaceae Prévot 1933* in *Bergey's manual of determinative bacteriology* edn. 6 (eds. Buchanan R.E.) 295-303 (The Williams & Wilkins Co., Baltimore, 1948)

129. Nabi, I.R. and Le, P.U. (2003). Caveolae/raft-dependent endocytosis. *J.Cell Biol.* **161** (4): 673-677.

130. Nagy, P., Vereb, G., Sebestyen, Z., Horvath, G., Lockett, S.J., Damjanovich, S., Park, J.W., Jovin, T.M., and Szollosi, J. (2002). Lipid rafts and the local density of ErbB proteins influence the biological role of homo- and heteroassociations of ErbB2. *J.Cell Sci.* **115** (Pt 22): 4251-4262.

131. Nassif, X. (1999). Interaction mechanisms of encapsulated meningococci with eucaryotic cells: what does this tell us about the crossing of the blood-brain barrier by Neisseria meningitidis? *Curr.Opin.Microbiol.* **2** (1): 71-77.

132. Nassif, X., Lowy, J., Stenberg, P., O'Gaora, P., Ganji, A., and So, M. (1993). Antigenic variation of pilin regulates adhesion of Neisseria meningitidis to human epithelial cells. *Mol.Microbiol.* **8** (4): 719-725.

133. Neame, S.J., Uff, C.R., Sheikh, H., Wheatley, S.C., and Isacke, C.M. (1995). CD44 exhibits a cell type dependent interaction with triton X-100 insoluble, lipid rich, plasma membrane domains. *J.Cell Sci.* **108** (Pt 9) 3127-3135.

134. Nichols, B.J., Kenworthy, A.K., Polishchuk, R.S., Lodge, R., Roberts, T.H., Hirschberg, K., Phair, R.D., and Lippincott-Schwartz, J. (2001). Rapid cycling of lipid raft markers between the cell surface and Golgi complex. *J.Cell Biol.* **153** (3): 529-541.

135. Nichols, B.J. and Lippincott-Schwartz, J. (2001). Endocytosis without clathrin coats. *Trends Cell Biol.* **11** (10): 406-412.

136. Niebuhr, K., Chakraborty, T., Rohde, M., Gazlig, T., Jansen, B., Kollner, P., and Wehland, J. (1993). Localization of the ActA polypeptide of Listeria monocytogenes in infected tissue culture cell lines: ActA is not associated with actin "comets". *Infect.Immun.* **61** (7): 2793-2802.

137. Niebuhr, K., Ebel, F., Frank, R., Reinhard, M., Domann, E., Carl, U.D., Walter, U., Gertler, F.B., Wehland, J., and Chakraborty, T. (1997). A novel proline-rich motif present in ActA of Listeria monocytogenes and cytoskeletal proteins is the ligand for the EVH1 domain, a protein module present in the Ena/VASP family. *EMBO J.* **16** (17): 5433-5444.

138. Norkin, L.C., Anderson, H.A., Wolfrom, S.A., and Oppenheim, A. (2002). Caveolar endocytosis of simian virus 40 is followed by brefeldin A-sensitive transport to the endoplasmic reticulum, where the virus disassembles. *J.Virol.* **76** (10): 5156-5166.

139. Oh, P., McIntosh, D.P., and Schnitzer, J.E. (1998). Dynamin at the neck of caveolae mediates their budding to form transport vesicles by GTP-driven fission from the plasma membrane of endothelium. *J.Cell Biol.* **141** (1): 101-114.

140. Orlandi, P.A. and Fishman, P.H. (1998). Filipin-dependent inhibition of cholera toxin: evidence for toxin internalization and activation through caveolae-like domains. *J.Cell Biol.* **141** (4): 905-915.

References

141. Orlichenko, L., Huang, B., Krueger, E., and McNiven, M.A. (2006). Epithelial growth factor-induced phosphorylation of caveolin 1 at tyrosine 14 stimulates caveolae formation in epithelial cells. *J.Biol.Chem.* **281** (8): 4570-4579.

142. Orth, J.D., Krueger, E.W., Cao, H., and McNiven, M.A. (2002). The large GTPase dynamin regulates actin comet formation and movement in living cells. *Proc.Natl.Acad.Sci.U.S.A* **99** (1): 167-172.

143. Ostrom, R.S., Liu, X., Head, B.P., Gregorian, C., Seasholtz, T.M., and Insel, P.A. (2002). Localization of adenylyl cyclase isoforms and G protein-coupled receptors in vascular smooth muscle cells: expression in caveolin-rich and noncaveolin domains. *Mol.Pharmacol.* **62** (5): 983-992.

144. Palade, G.E. (1953). Fine structure of blood capillaries. *J.Appl.Physiol* **24** 1424-

145. Parton, R.G., Joggerst, B., and Simons, K. (1994). Regulated internalization of caveolae. *J.Cell Biol.* **127** (5): 1199-1215.

146. Paruchuri, D.K., Seifert, H.S., Ajioka, R.S., Karlsson, K.A., and So, M. (1990). Identification and characterization of a Neisseria gonorrhoeae gene encoding a glycolipid-binding adhesin. *Proc.Natl.Acad.Sci.U.S.A* **87** (1): 333-337.

147. Pelkmans, L., Burli, T., Zerial, M., and Helenius, A. (2004). Caveolin-stabilized membrane domains as multifunctional transport and sorting devices in endocytic membrane traffic. *Cell* **118** (6): 767-780.

148. Pelkmans, L., Kartenbeck, J., and Helenius, A. (2001). Caveolar endocytosis of simian virus 40 reveals a new two-step vesicular-transport pathway to the ER. *Nat.Cell Biol.* **3** (5): 473-483.

149. Pelkmans, L., Puntener, D., and Helenius, A. (2002). Local actin polymerization and dynamin recruitment in SV40-induced internalization of caveolae. *Science* **296** (5567): 535-539.

150. Pelkmans, L. and Zerial, M. (2005). Kinase-regulated quantal assemblies and kiss-and-run recycling of caveolae. *Nature* **436** (7047): 128-133.

151. Pike, L.J. (2005). Growth factor receptors, lipid rafts and caveolae: an evolving story. *Biochim.Biophys.Acta* **1746** (3): 260-273.

152. Porat, N., Apicella, M.A., and Blake, M.S. (1995). Neisseria gonorrhoeae utilizes and enhances the biosynthesis of the asialoglycoprotein receptor expressed on the surface of the hepatic HepG2 cell line. *Infect.Immun.* **63** (4): 1498-1506.

153. Preston, A., Mandrell, R.E., Gibson, B.W., and Apicella, M.A. (1996). The lipooligosaccharides of pathogenic gram-negative bacteria. *Crit Rev.Microbiol.* **22** (3): 139-180.

154. Pujol, C., Eugene, E., de Saint, M.L., and Nassif, X. (1997). Interaction of Neisseria meningitidis with a polarized monolayer of epithelial cells. *Infect.Immun.* **65** (11): 4836-4842.

155. Pujol, C., Eugene, E., Marceau, M., and Nassif, X. (1999). The meningococcal PilT protein is required for induction of intimate attachment to epithelial cells following pilus-mediated adhesion. *Proc.Natl.Acad.Sci.U.S.A* **96** (7): 4017-4022.

156. Rajalingam, K., Sharma, M., Paland, N., Hurwitz, R., Thieck, O., Oswald, M., Machuy, N., and Rudel, T. (2006). IAP-IAP complexes required for apoptosis resistance of C. trachomatis-infected cells. *PLoS.Pathog.* **2** (10): e114-

157. Ramsey, K.H., SCHNEIDER, H., Cross, A.S., Boslego, J.W., Hoover, D.L., Staley, T.L., Kuschner, R.A., and Deal, C.D. (1995). Inflammatory cytokines produced in response to experimental human gonorrhea. *J.Infect.Dis.* **172** (1): 186-191.

References

158. Razani, B., Combs, T.P., Wang, X.B., Frank, P.G., Park, D.S., Russell, R.G., Li, M., Tang, B., Jelicks, L.A., Scherer, P.E., and Lisanti, M.P. (2002). Caveolin-1-deficient mice are lean, resistant to diet-induced obesity, and show hypertriglyceridemia with adipocyte abnormalities. *J.Biol.Chem.* **277** (10): 8635-8647.

159. Razani, B. and Lisanti, M.P. (2001). Two distinct caveolin-1 domains mediate the functional interaction of caveolin-1 with protein kinase A. *Am.J.Physiol Cell Physiol* **281** (4): C1241-C1250.

160. Razani, B., Rubin, C.S., and Lisanti, M.P. (1999). Regulation of cAMP-mediated signal transduction via interaction of caveolins with the catalytic subunit of protein kinase A. *J.Biol.Chem.* **274** (37): 26353-26360.

161. Rechner, C., Kuhlewein, C., Muller, A., Schild, H., and Rudel, T. (2007). Host glycoprotein Gp96 and scavenger receptor SREC interact with PorB of disseminating Neisseria gonorrhoeae in an epithelial invasion pathway. *Cell Host.Microbe* **2** (6): 393-403.

162. Reinhard, M., Giehl, K., Abel, K., Haffner, C., Jarchau, T., Hoppe, V., Jockusch, B.M., and Walter, U. (1995a). The proline-rich focal adhesion and microfilament protein VASP is a ligand for profilins. *EMBO J.* **14** (8): 1583-1589.

163. Reinhard, M., Jarchau, T., and Walter, U. (2001). Actin-based motility: stop and go with Ena/VASP proteins. *Trends Biochem.Sci.* **26** (4): 243-249.

164. Reinhard, M., Jouvenal, K., Tripier, D., and Walter, U. (1995b). Identification, purification, and characterization of a zyxin-related protein that binds the focal adhesion and microfilament protein VASP (vasodilator-stimulated phosphoprotein). *Proc.Natl.Acad.Sci.U.S.A* **92** (17): 7956-7960.

165. Rothberg, K.G., Heuser, J.E., Donzell, W.C., Ying, Y.S., Glenney, J.R., and Anderson, R.G. (1992). Caveolin, a protein component of caveolae membrane coats. *Cell* **68** (4): 673-682.

166. Roy, N., Deveraux, Q.L., Takahashi, R., Salvesen, G.S., and Reed, J.C. (1997). The c-IAP-1 and c-IAP-2 proteins are direct inhibitors of specific caspases. *EMBO J.* **16** (23): 6914-6925.

167. Rudel, T., Scheurerpflug, I., and Meyer, T.F. (1995). Neisseria PilC protein identified as type-4 pilus tip-located adhesin. *Nature* **373** (6512): 357-359.

168. Rudel, T., Schmid, A., Benz, R., Kolb, H.A., Lang, F., and Meyer, T.F. (1996). Modulation of Neisseria porin (PorB) by cytosolic ATP/GTP of target cells: parallels between pathogen accommodation and mitochondrial endosymbiosis. *Cell* **85** (3): 391-402.

169. Sainio, M., Zhao, F., Heiska, L., Turunen, O., den Bakker, M., Zwarthoff, E., Lutchman, M., Rouleau, G.A., Jaaskelainen, J., Vaheri, A., and Carpen, O. (1997). Neurofibromatosis 2 tumor suppressor protein colocalizes with ezrin and CD44 and associates with actin-containing cytoskeleton. *J.Cell Sci.* **110 (Pt 18)** 2249-2260.

170. Sargiacomo, M., Scherer, P.E., Tang, Z., Kubler, E., Song, K.S., Sanders, M.C., and Lisanti, M.P. (1995). Oligomeric structure of caveolin: implications for caveolae membrane organization. *Proc.Natl.Acad.Sci.U.S.A* **92** (20): 9407-9411.

171. Sargiacomo, M., Sudol, M., Tang, Z., and Lisanti, M.P. (1993). Signal transducing molecules and glycosyl-phosphatidylinositol-linked proteins form a caveolin-rich insoluble complex in MDCK cells. *J.Cell Biol.* **122** (4): 789-807.

172. Sayeed, Z.A., Bhaduri, U., Howell, E., and Meyers, H.L., Jr. (1972). Gonococcal meningitis. A review. *JAMA* **219** (13): 1730-1731.

References

173. Scherer, P.E., Okamoto, T., Chun, M., Nishimoto, I., Lodish, H.F., and Lisanti, M.P. (1996). Identification, sequence, and expression of caveolin-2 defines a caveolin gene family. *Proc.Natl.Acad.Sci.U.S.A* **93** (1): 131-135.

174. Scherer, P.E., Tang, Z., Chun, M., Sargiacomo, M., Lodish, H.F., and Lisanti, M.P. (1995). Caveolin isoforms differ in their N-terminal protein sequence and subcellular distribution. Identification and epitope mapping of an isoform-specific monoclonal antibody probe. *J.Biol.Chem.* **270** (27): 16395-16401.

175. Scheuerpflug, I., Rudel, T., Ryll, R., Pandit, J., and Meyer, T.F. (1999). Roles of PilC and PilE proteins in pilus-mediated adherence of Neisseria gonorrhoeae and Neisseria meningitidis to human erythrocytes and endothelial and epithelial cells. *Infect.Immun.* **67** (2): 834-843.

176. Schroeder, R., London, E., and Brown, D. (1994). Interactions between saturated acyl chains confer detergent resistance on lipids and glycosylphosphatidylinositol (GPI)-anchored proteins: GPI-anchored proteins in liposomes and cells show similar behavior. *Proc.Natl.Acad.Sci.U.S.A* **91** (25): 12130-12134.

177. Schroeder, R.J., Ahmed, S.N., Zhu, Y., London, E., and Brown, D.A. (1998). Cholesterol and sphingolipid enhance the Triton X-100 insolubility of glycosylphosphatidylinositol-anchored proteins by promoting the formation of detergent-insoluble ordered membrane domains. *J.Biol.Chem.* **273** (2): 1150-1157.

178. Schwencke, C., Okumura, S., Yamamoto, M., Geng, Y.J., and Ishikawa, Y. (1999). Colocalization of beta-adrenergic receptors and caveolin within the plasma membrane. *J.Cell Biochem.* **75** (1): 64-72.

179. Seifert, H.S., Wright, C.J., Jerse, A.E., Cohen, M.S., and Cannon, J.G. (1994). Multiple gonococcal pilin antigenic variants are produced during experimental human infections. *J.Clin.Invest* **93** (6): 2744-2749.

180. Serino, L., Nesta, B., Leuzzi, R., Fontana, M.R., Monaci, E., Mocca, B.T., Cartocci, E., Masignani, V., Jerse, A.E., Rappuoli, R., and Pizza, M. (2007). Identification of a new OmpA-like protein in Neisseria gonorrhoeae involved in the binding to human epithelial cells and in vivo colonization. *Mol.Microbiol.* **64** (5): 1391-1403.

181. Shen, Y., Naujokas, M., Park, M., and Ireton, K. (2000). InlB-dependent internalization of Listeria is mediated by the Met receptor tyrosine kinase. *Cell* **103** (3): 501-510.

182. Shin, J.S., Gao, Z., and Abraham, S.N. (2000). Involvement of cellular caveolae in bacterial entry into mast cells. *Science* **289** (5480): 785-788.

183. Sieveking, D., Mitchell, H.M., and Day, A.S. (2004). Gastric epithelial cell CXC chemokine secretion following Helicobacter pylori infection in vitro. *J.Gastroenterol.Hepatol.* **19** (9): 982-987.

184. Simons, K. and Ikonen, E. (1997). Functional rafts in cell membranes. *Nature* **387** (6633): 569-572.

185. Singer, S.J. and Nicolson, G.L. (1972). The fluid mosaic model of the structure of cell membranes. *Science* **175** (23): 720-731.

186. Skoudy, A., Mounier, J., Aruffo, A., Ohayon, H., Gounon, P., Sansonetti, P., and Tran, V.N. (2000). CD44 binds to the Shigella IpaB protein and participates in bacterial invasion of epithelial cells. *Cell Microbiol.* **2** (1): 19-33.

187. Smith, H., Cole, J.A., and Parsons, N.J. (1992). The sialylation of gonococcal lipopolysaccharide by host factors: a major impact on pathogenicity. *FEMS Microbiol.Lett.* **79** (1-3): 287-292.

References

188. Smith, H., Parsons, N.J., and Cole, J.A. (1995). Sialylation of neisserial lipopolysaccharide: a major influence on pathogenicity. *Microb.Pathog.* **19** (6): 365-377.

189. Smith, K.E., Gu, C., Fagan, K.A., Hu, B., and Cooper, D.M. (2002). Residence of adenylyl cyclase type 8 in caveolae is necessary but not sufficient for regulation by capacitative Ca(2+) entry. *J.Biol.Chem.* **277** (8): 6025-6031.

190. Song, W., Ma, L., Chen, R., and Stein, D.C. (2000). Role of lipooligosaccharide in Opa-independent invasion of Neisseria gonorrhoeae into human epithelial cells. *J.Exp.Med.* **191** (6): 949-960.

191. Stahlhut, M. and van Deurs, B. (2000). Identification of filamin as a novel ligand for caveolin-1: evidence for the organization of caveolin-1-associated membrane domains by the actin cytoskeleton. *Mol.Biol.Cell* **11** (1): 325-337.

192. Stang, E., Kartenbeck, J., and Parton, R.G. (1997). Major histocompatibility complex class I molecules mediate association of SV40 with caveolae. *Mol.Biol.Cell* **8** (1): 47-57.

193. Swanson, J. (1978). Studies on gonococcus infection. XII. Colony color and opacity varienats of gonococci. *Infect.Immun.* **19** (1): 320-331.

194. Swanson, J., Robbins, K., Barrera, O., Corwin, D., Boslego, J., Ciak, J., Blake, M., and Koomey, J.M. (1987). Gonococcal pilin variants in experimental gonorrhea. *J.Exp.Med.* **165** (5): 1344-1357.

195. Swanson, K.V., Jarvis, G.A., Brooks, G.F., Barham, B.J., Cooper, M.D., and Griffiss, J.M. (2001). CEACAM is not necessary for Neisseria gonorrhoeae to adhere to and invade female genital epithelial cells. *Cell Microbiol.* **3** (10): 681-691.

196. Sweet, R.L. (1987). Sexually transmitted diseases. Pelvic inflammatory disease and infertility in women. *Infect.Dis.Clin.North Am.* **1** (1): 199-215.

197. Tang, Z., Scherer, P.E., Okamoto, T., Song, K., Chu, C., Kohtz, D.S., Nishimoto, I., Lodish, H.F., and Lisanti, M.P. (1996). Molecular cloning of caveolin-3, a novel member of the caveolin gene family expressed predominantly in muscle. *J.Biol.Chem.* **271** (4): 2255-2261.

198. Taylor, R.K., Miller, V.L., Furlong, D.B., and Mekalanos, J.J. (1987). Use of phoA gene fusions to identify a pilus colonization factor coordinately regulated with cholera toxin. *Proc.Natl.Acad.Sci.U.S.A* **84** (9): 2833-2837.

199. Thomsen, P., Roepstorff, K., Stahlhut, M., and van Deurs, B. (2002). Caveolae are highly immobile plasma membrane microdomains, which are not involved in constitutive endocytic trafficking. *Mol.Biol.Cell* **13** (1): 238-250.

200. Thorn, H., Stenkula, K.G., Karlsson, M., Ortegren, U., Nystrom, F.H., Gustavsson, J., and Stralfors, P. (2003). Cell surface orifices of caveolae and localization of caveolin to the necks of caveolae in adipocytes. *Mol.Biol.Cell* **14** (10): 3967-3976.

201. Tilney, L.G. and Portnoy, D.A. (1989). Actin filaments and the growth, movement, and spread of the intracellular bacterial parasite, Listeria monocytogenes. *J.Cell Biol.* **109** (4 Pt 1): 1597-1608.

202. Toleman, M., Aho, E., and Virji, M. (2001). Expression of pathogen-like Opa adhesins in commensal Neisseria: genetic and functional analysis. *Cell Microbiol.* **3** (1): 33-44.

203. Towbin, H., Staehelin, T., and Gordon, J. (1979). Electrophoretic transfer of proteins from polyacrylamide gels to nitrocellulose sheets: procedure and some applications. *Proc.Natl.Acad.Sci.U.S.A* **76** (9): 4350-4354.

References

204. Toya, Y., Schwencke, C., Couet, J., Lisanti, M.P., and Ishikawa, Y. (1998). Inhibition of adenylyl cyclase by caveolin peptides. *Endocrinology* **139** (4): 2025-2031.

205. Trasak, C., Zenner, G., Vogel, A., Yuksekdag, G., Rost, R., Haase, I., Fischer, M., Israel, L., Imhof, A., Linder, S., Schleicher, M., and Aepfelbacher, M. (2007). Yersinia protein kinase YopO is activated by a novel G-actin binding process. *J.Biol.Chem.* **282** (4): 2268-2277.

206. Unsworth, K.E., Way, M., McNiven, M., Machesky, L., and Holden, D.W. (2004). Analysis of the mechanisms of Salmonella-induced actin assembly during invasion of host cells and intracellular replication. *Cell Microbiol.* **6** (11): 1041-1055.

207. van Putten, J.P. (1993). Phase variation of lipopolysaccharide directs interconversion of invasive and immuno-resistant phenotypes of Neisseria gonorrhoeae. *EMBO J.* **12** (11): 4043-4051.

208. van Putten, J.P., Duensing, T.D., and Carlson, J. (1998a). Gonococcal invasion of epithelial cells driven by P.IA, a bacterial ion channel with GTP binding properties. *J.Exp.Med.* **188** (5): 941-952.

209. van Putten, J.P., Duensing, T.D., and Cole, R.L. (1998b). Entry of OpaA+ gonococci into HEp-2 cells requires concerted action of glycosaminoglycans, fibronectin and integrin receptors. *Mol.Microbiol.* **29** (1): 369-379.

210. Virji, M., Makepeace, K., Peak, I.R., Ferguson, D.J., Jennings, M.P., and Moxon, E.R. (1995). Opc- and pilus-dependent interactions of meningococci with human endothelial cells: molecular mechanisms and modulation by surface polysaccharides. *Mol.Microbiol.* **18** (4): 741-754.

211. Vogel, U., Elias, J., and Hellenbrand, W. (2007). Invasive Meningokokken-Erkrankungen im Jahr 2006. *Epidemiologisches Bulletin*(32): 297-306.

212. Vogel, U. and Frosch, M. (1999). Mechanisms of neisserial serum resistance. *Mol.Microbiol.* **32** (6): 1133-1139.

213. Voon, D.C., Subrata, L.S., Karimi, M., Ulgiati, D., and Abraham, L.J. (2004). TNF and phorbol esters induce lymphotoxin-beta expression through distinct pathways involving Ets and NF-kappa B family members. *J.Immunol.* **172** (7): 4332-4341.

214. Walders-Harbeck, B., Khaitlina, S.Y., Hinssen, H., Jockusch, B.M., and Illenberger, S. (2002). The vasodilator-stimulated phosphoprotein promotes actin polymerisation through direct binding to monomeric actin. *FEBS Lett.* **529** (2-3): 275-280.

215. Wang, D., Wang, H., Brown, J., Daikoku, T., Ning, W., Shi, Q., Richmond, A., Strieter, R., Dey, S.K., and DuBois, R.N. (2006). CXCL1 induced by prostaglandin E2 promotes angiogenesis in colorectal cancer. *J.Exp.Med.* **203** (4): 941-951.

216. Wang, J., Gray-Owen, S.D., Knorre, A., Meyer, T.F., and Dehio, C. (1998). Opa binding to cellular CD66 receptors mediates the transcellular traversal of Neisseria gonorrhoeae across polarized T84 epithelial cell monolayers. *Mol.Microbiol.* **30** (3): 657-671.

217. Ward, M.E. and Murray, A. (1984). Control mechanisms governing the infectivity of Chlamydia trachomatis for HeLa cells: mechanisms of endocytosis. *J.Gen.Microbiol.* **130** (7): 1765-1780.

218. Watson, R.T., Shigematsu, S., Chiang, S.H., Mora, S., Kanzaki, M., Macara, I.G., Saltiel, A.R., and Pessin, J.E. (2001). Lipid raft microdomain compartmentalization of TC10 is required for insulin signaling and GLUT4 translocation. *J.Cell Biol.* **154** (4): 829-840.

219. Weel, J.F. and van Putten, J.P. (1991). Fate of the major outer membrane protein P.IA in early and late events of gonococcal infection of epithelial cells. *Res.Microbiol.* **142** (9): 985-993.

220. Welch, M.D., Iwamatsu, A., and Mitchison, T.J. (1997). Actin polymerization is induced by Arp2/3 protein complex at the surface of Listeria monocytogenes. *Nature* **385** (6613): 265-269.

221. Wen, K.K., Giardina, P.C., Blake, M.S., Edwards, J., Apicella, M.A., and Rubenstein, P.A. (2000). Interaction of the gonococcal porin P.IB with G- and F-actin. *Biochemistry* **39** (29): 8638-8647.

222. Whitchurch, C.B., Hobbs, M., Livingston, S.P., Krishnapillai, V., and Mattick, J.S. (1991). Characterisation of a Pseudomonas aeruginosa twitching motility gene and evidence for a specialised protein export system widespread in eubacteria. *Gene* **101** (1): 33-44.

223. Wiley, D.J., Nordfeldth, R., Rosenzweig, J., DaFonseca, C.J., Gustin, R., Wolf-Watz, H., and Schesser, K. (2006). The Ser/Thr kinase activity of the Yersinia protein kinase A (YpkA) is necessary for full virulence in the mouse, mollifying phagocytes, and disrupting the eukaryotic cytoskeleton. *Microb.Pathog.* **40** (5): 234-243.

224. Williams, T.M., Cheung, M.W., Park, D.S., Razani, B., Cohen, A.W., Muller, W.J., Di Vizio, D., Chopra, N.G., Pestell, R.G., and Lisanti, M.P. (2003). Loss of caveolin-1 gene expression accelerates the development of dysplastic mammary lesions in tumor-prone transgenic mice. *Mol.Biol.Cell* **14** (3): 1027-1042.

225. Wolfgang, M., van Putten, J.P., Hayes, S.F., Dorward, D., and Koomey, M. (2000). Components and dynamics of fiber formation define a ubiquitous biogenesis pathway for bacterial pili. *EMBO J.* **19** (23): 6408-6418.

226. Wooldridge, K.G., Williams, P.H., and Ketley, J.M. (1996). Host signal transduction and endocytosis of Campylobacter jejuni. *Microb.Pathog.* **21** (4): 299-305.

227. YAMADA, E. (1955). The fine structure of the gall bladder epithelium of the mouse. *J.Biophys.Biochem.Cytol.* **1** (5): 445-458.

228. Yamasaki, R., Bacon, B.E., Nasholds, W., SCHNEIDER, H., and Griffiss, J.M. (1991). Structural determination of oligosaccharides derived from lipooligosaccharide of Neisseria gonorrhoeae F62 by chemical, enzymatic, and two-dimensional NMR methods. *Biochemistry* **30** (43): 10566-10575.

229. Yazdankhah, S.P., Kriz, P., Tzanakaki, G., Kremastinou, J., Kalmusova, J., Musilek, M., Alvestad, T., Jolley, K.A., Wilson, D.J., McCarthy, N.D., Caugant, D.A., and Maiden, M.C. (2004). Distribution of serogroups and genotypes among disease-associated and carried isolates of Neisseria meningitidis from the Czech Republic, Greece, and Norway. *J.Clin.Microbiol.* **42** (11): 5146-5153.

230. Zaas, D.W., Duncan, M.J., Li, G., Wright, J.R., and Abraham, S.N. (2005). Pseudomonas invasion of type I pneumocytes is dependent on the expression and phosphorylation of caveolin-2. *J.Biol.Chem.* **280** (6): 4864-4872.

231. Zhang, X.L., Tsui, I.S., Yip, C.M., Fung, A.W., Wong, D.K., Dai, X., Yang, Y., Hackett, J., and Morris, C. (2000). Salmonella enterica serovar typhi uses type IVB pili to enter human intestinal epithelial cells. *Infect.Immun.* **68** (6): 3067-3073.

10 Index

10.1 Figure Index

Figure 2-1: Neisserial factors involved in adhesion and invasion 15
Figure 2-2: Model for *N. meningitides* invasion 22
Figure 2-3: Model for caveolae endocytosis 29
Figure 4-1: Expression of caveolin-1 in epithelial cell lines 33
Table 4-1: Gene upregulation in *N.gonorrhoeae* infected AGS-pcDNA3 cells 35
Table 4-2: Gene upregulation in *N. gonorrhoeae* infected AGS-179 cells 36
Figure 4-2: Fold change of PKA-RIβ mRNA after *Ngo* infection in caveolin-expressing AGS and AGS control cells 37
Figure 4-3: Influence of the PKA agonist forskolin and antagonist H-89 on *Ngo* internalization 38
Figure 4-4: Increase in *Ngo* internalization due to PKA inhibition is pilus-dependent 40
Figure 4-5: Activation of the AC/PKA pathway upon *N. gonorrhoeae* infection 42
Figure 4-6: VASP recruitment to *N. gonorrhoeae* in VASP-GFP transfected ME180 cells 43
Figure 4-7: Influence of VASP and PKA-RIβ down-regulation in ME180 cells on *N. gonorrhoeae* internalization 45
Figure 4-8: Influence of VASP and PKA-RIβ down-regulation in AGS cells on *N. gonorrhoeae* internalization 46
Figure 4-9: Influence of actin cytoskeleton disruption on *N. gonorrhoeae* invasion in VASP knockdown cells 47
Figure 4-10: Interference with the AC/PKA pathway affects caveolin recruitment 48
Figure 4-11: Survival of intracellular *N. gonorrhoeae* is prolonged due to AC/PKA inhibition 50
Figure 4-12: AC/PKA pathway inhibition blocks association of intracellular *N. gonorrhoeae* with lysosomes 51
Figure 4-13: ErbB2 recruitment to gonococcal microcolonies 52
Figure 4-14: Association of pilus subunit PilC with ErbB2 53
Figure 4-15: Src activity in *N. gonorrhoeae* infected and pharmacologically treated cells 54
Figure 4-16: Actin recruitment of piliated *N. gonorrhoeae* in infected HeLa cells 55

Index

Figure 4-17: Cortical plaque formation in epithelial cells infected with *N. gonorrhoeae* 56

Figure 4-18: Actin comets emanate from *N. gonorrhoaea* triggered actin plaques... 57

Figure 4-19: VASP comets in epithelial cells 58

Figure 5-1: Inhibition of *N. gonorrhoeae* internalization into epithelial cells 65

Figure 5-2: Model for *N. gonorrhoeae* induced caveolin- and actin clustering 68

10.2 Abbreviations

General abbreviations	
Å	Ångstrom
A/E lesions	attaching and effacing lesions
cAMP	cyclic adenosine monophosphate
CCE	capacitive calcium entry
C-MAD	COOH-terminal attachment domain
cyt D	cytochalasin D
ddH$_2$0	double distilled water
DNA	desoxyribonucleic acid
ECL	enhanced chemoluminescence
EDTA	ethylenediaminetetraacetic acid
ER	endoplasmic reticulum
ERGIC	ER/Golgi intermediate compartment
GPI	glycosylphosphatidyl-inositol
GTP	guanosine triphosphate
IB	immunoblot
IF	immunofluorescence
Lat A	latrunculin A
LDAO	lauryldimethylamine-oxide
MOI	multiplicity of infection
N-MAD	NH$_2$-terminal attachment domain
P$^+$/P$^-$	piliated/non-piliated
PFA	para-formaldehyde
PVDF	poly-venylidene-fluoride
PBS	phosphate buffered saline
qRT-PCR	quantitative real time polymerase chain reaction
RNA	ribonucleic acid
RNAi	RNA interference
siRNA	small interfering RNA
T3SS	type 3 secretion system
TBS	Tris buffered saline
TBS-T	TBS added with Tween20

Index

General abbreviations

TE buffer	Tris-EDTA buffer
Tfp	type IV pili
Tris	tris(hydroxymethyl)aminomethane
w/o	without

Abbreviations of proteins and genes

ASPG-R	asiologlycoprotein receptor
BSA	bovine serum albumin
Cav1	caveolin-1
CD46	cluster of differentiation 46
Cdc42	cell division cycle 42
CEACAM	carcinoembryonic cell adhesion molecule
CMP-NANA	cytidinemonophosphate-N-acetylneuraminic acid
cyt D	cytochalasin D
DGI	disseminated gonococcal disease
EGF	epidermal growth factor
EGFR	epidermal growth factor receptor
ErB2	erythroblastosis oncogne B2
EspF$_U$	EPEC secreted protein F$_U$
GFP	green fluorescent protein
Gp96	glycoprotein 96
GAP-DH	glyceraldehyde-3-dehydrogenase
Gro-alpha	growth-regulated oncogene alpha
HSPG	heperansulphate proteoglycan
ICAM-1	intercellular cell adhesion molecule 1
I-domain	inserted domain
IgA	immunoglobulin A
Inl A/B	internalin A/B
Ipa B	invasion plasmid antigen B
LAMP1	lysosomal-associated membrane proteine 1
LOS	lipooligosaccharide
MCP	membrane cofactor receptor

Abbreviations of proteins and genes

MHC	major histocompatibility complex
Opa$_{CEA}$	CEACAM-recognizing colony opacity associated protein
Opa$_{HS}$	HSPG-recognizing Opa
PID	pelvic inflammatory diesease
pilC/D/E etc.	pilus subunits
pilS	pilE pseudogene
PorA/B	porin A/B
Rac	Ras-related C3botulinum toxin substrate
Rho	Ras homolog gene family
ROKα	Rho kinase alpha
Src	sarcoma
TccP	Tir-cytoskeleton coupling protein
Tir	translocated intimin receptor
TNFα	tumor necrosis factor alpha
VASP	vasodilator stimulated phosphoprotein

Abbreviations of organisms

E. coli	Escherichia coli
EHEC	enterohemorrhagic *E. coli*
EPEC	enteropathogenic *E. coli*
Ngo	*Neisseria gonorrhoeae*
Nmg	*Neisseria meningitides*
SV40	simian virus 40

10.3 Acknowledgements

I would like to thank Prof. Dr. Thomas F. Meyer for the supervision of this thesis, for providing this interesting project and for the warm reception in his working group.

Thanks to Prof. Dr. Petra Knaus for supervising this thesis as a second reviewer who did not hesitate to take over the job.

Many thanks to Jura Churin who pushed me through the time as a PhD student in the lab with his brilliant ideas which significally influenced the focus of my work. Without him, the PKA story would never have started. In addition, he not only was the mind behind the project, he also contributed a lot to the daily work by helping at the bench an by controversely discussing results. It was a great privilege to have him available around the clock in the lab.

I also would like to thank Marieluise Kirchner for introducing me to the *Neisseria* field and for her encouraging help with the labwork, especially for her assistance with the PilC purification. Her microarray experiments with caveolin-1 expressing AGS cells lay the basis for my work.

A bunch of thanks goes to the people of the *Neisseria* group and the people of office 2.38, namely Manuela Dietrich, Rebekka Munke, Manuel Koch, Oliver Riede and Piet Böttcher. Working with you made even bad days bearable. Thanks to Hans Thorn for the help with the life cell imaging microscope. Many thanks to Dr. Yoshan Moodley and Franziska Grzegorzewski for proofreading the manuscript.

A very sincere thanks goes, once again, to my parents, Rosemarie and Walter de Graaf. It is hard to imagine that this project would have have been possible without your support.

Many thanks to Christina Zech. Without her the project would never have come to an end.

i want morebooks!

Buy your books fast and straightforward online - at one of world's fastest growing online book stores! Environmentally sound due to Print-on-Demand technologies.

Buy your books online at
www.get-morebooks.com

Kaufen Sie Ihre Bücher schnell und unkompliziert online – auf einer der am schnellsten wachsenden Buchhandelsplattformen weltweit! Dank Print-On-Demand umwelt- und ressourcenschonend produziert.

Bücher schneller online kaufen
www.morebooks.de

VDM Verlagsservicegesellschaft mbH
Heinrich-Böcking-Str. 6-8   Telefon: +49 681 3720 174   info@vdm-vsg.de
D - 66121 Saarbrücken   Telefax: +49 681 3720 1749   www.vdm-vsg.de

Printed by Books on Demand GmbH, Norderstedt / Germany